INTERNATIONAL DEVELOPMENT IN FOCUS

Getting Down to Earth

Are Satellites Reliable for Measuring
Air Pollutants That Cause Mortality in
Low- and Middle-Income Countries?

Contents

Preface   *ix*
About This Work   *xi*
Acknowledgments   *xiii*
Executive Summary   *xv*
Abbreviations   *xxi*
Glossary   *xxiii*

CHAPTER 1   **Contextual Background and Objectives**   **1**

CHAPTER 2   **Introduction**   **3**
Report methodology   8
Outline of this report   9
Note   9
References   9

CHAPTER 3   **Literature Review of Approaches Used to Combine Satellite Observations with Ground-Level Monitoring in Urban Areas**   **11**
Retrieval of AOD from satellite observations of reflected solar radiation   12
The translation of AOD to ground-level $PM_{2.5}$ concentrations   16
Combining satellite $PM_{2.5}$ estimates and GLM data into high-resolution grids centered on selected cities   18
Notes   18
References   19

CHAPTER 4   **Quality-Assurance Procedures in Low- and Middle-Income Countries**   **21**
Issues affecting the availability and accuracy of GLM data in LMICs   21
Using automated scripts to flag potentially incorrect GLM data   24
Examining metadata, warning files, and operator logs   25
Note   26

CHAPTER 5   **Using Satellite Data for Daily, City-Scale $PM_{2.5}$ Monitoring**   **27**
Initial testing of satellite products in selected cities   28
Identifying the type of cities where satellites work best   34
Notes   41
References   41

CHAPTER 6   **Conclusions and Recommendations**   **43**

Conclusions   44

Recommendations   47

Note   50

References   51

Appendix A   **Literature Review**   **53**

Appendix B   **Converting Satellite Aerosol Optical Depth to Ground-Level PM$_{2.5}$**   **97**

Appendix C   **Evaluation of Satellite Approaches**   **121**

Appendix D   **Quality Considerations for Ground-Level-Monitoring Data in Low- and Middle-Income Countries**   **149**

Box

4.1   Open-access web platform for air quality data   22

Figures

2.1   MODIS Aqua image of wildfire smoke being transported eastward to the Chesapeake Bay, US   4

2.2   Current and future coverage provided by available polar and geostationary weather satellites   5

2.3   How satellites can measure aerosol optical depth from reflected sunlight   6

2.4   Highly idealized example of three different aerosol vertical profiles, each with different ground-level PM$_{2.5}$ concentrations, that result in the same aerosol-optical-depth value of one when measured by satellites   7

3.1   Annual average ground-level PM$_{10}$ concentrations, London, UK, and Rome, Italy   12

3.2   MODIS Dark Target 10-kilometer and 3-kilometer aerosol-optical-depth products retrieved for clear land and ocean fields of view and the local 5-kilometer average derived from the products, outer circle, compared to ground-based measurements, inner circle, over Baltimore, US   14

3.3   Example of good-quality aerosol optical depth 550 nm observations from a geostationary satellite from the US GOES-R SMAOD product   15

3.4   Valid MODIS Terra aerosol-optical-depth retrievals and Lima, Peru, monitoring sites in the OpenAQ database, 2016–17   16

4.1   Example of US federal equivalent method measurement and uncertainties: Data set slope, intercept, and limits   23

5.1   Comparison of the chemical-transport-model–based estimates for daily average ground-level PM$_{2.5}$ concentrations in micrograms per cubic meter over Delhi, India, using VIIRS and MODIS Terra, November 1, 2017   32

5.2   Comparison of the chemical-transport-model–based estimates for daily average ground-level PM$_{2.5}$ concentrations in micrograms per cubic meter over Accra, Ghana, using SEVIRI and MODIS Aqua, December 26, 2017   33

5.3   Correlation coefficient R for the statistical and chemical-transport-model–based methods for the different low- and middle-income country cities   37

5.4   Mean normalized bias for the statistical and chemical-transport-model–based methods for the different low- and middle-income country cities   37

5.5   Mean normalized gross error for the statistical and chemical-transport-model–based methods for the different low- and middle-income country cities   38

A.1   Coverage provided by available polar and geostationary weather satellites   57

A.2   MODIS Dark Target 10-kilometer and 3-kilometer aerosol-optical-depth products retrieved for clear land and ocean fields of view and the local 5-kilometer average derived from the products, outer circle, compared to ground-based measurements, inner circle, over Baltimore, US   60

A.3 Comparison of VIIRS 750-meter and MAIAC one-kilometer aerosol-optical-depth products during western US fires, 2013   63

A.4 Detection of temporal variations in aerosol optical depth with the SEVIRI-MSG aerosol over land product compared to AERONET measurements at Palaiseau, France, and to MODIS, July 14, 2006   66

A.5 Full disk coverage at 550 nanometers of GOES-R SMAOD product over ocean and land   67

A.6 Planetary boundary layer height in areas surrounding Washington, DC, US, on July 14, 2011, at 21:00 UTC   72

B.1 Illustration of the regions covered by a 1° by 1° box around each focus region   100

B.2 Local Ulaanbaatar, Mongolia, ground-level-monitoring network and US Diplomatic Post annual average $PM_{2.5}$ concentrations as well as a gridded estimate produced via ordinary kriging   105

B.3 Example of satellite coverage showing the CTM–based estimates for daily average ground-level $PM_{2.5}$ concentrations for all valid MODIS Terra and Aqua aerosol-optical-depth retrievals, Lima, Peru, May 12, 2017   107

B.4 Residual evaluation plots for Ulaanbaatar, Mongolia, using the MODIS Aqua aerosol-optical-depth product   108

B.5 Observed and GAM–predicted daily average $PM_{2.5}$ values for the US Diplomatic Posts in Delhi, India, and Ulaanbaatar, Mongolia, for both MODIS Aqua and Terra   109

B.6 Observed and CTM–based estimates daily average $PM_{2.5}$ values for the US Diplomatic Posts in Delhi, India, and Ulaanbaatar, Mongolia, for both MODIS Aqua and Terra   110

B.7 Monthly average $PM_{2.5}$ surface concentration, 2017   112

B.8 Observed values and VIIRS-based estimates of daily average $PM_{2.5}$ concentrations for the Income Tax Office in Delhi, India, using GAM and CTM–based approaches   114

B.9 A comparison of the CTM–based estimates for daily-average ground-level $PM_{2.5}$ concentrations over Delhi, India, using VIIRS and MODIS Terra, November 1, 2017   115

B.10 A comparison of the CTM–based estimates for daily-average ground-level $PM_{2.5}$ concentrations over Accra, Ghana, using SEVIRI and MODIS Aqua, December 26, 2017   116

C.1 Residual evaluation plots for the generalized additive model trained on the Terra satellite data, Delhi, India   123

C.2 Plot of the dependence of the ground-level $PM_{2.5}$ estimate from the generalized additive model on the ratio of the Terra aerosol optical depth and the MERRA planetary-boundary-layer height, Delhi, India   124

C.3 Scatterplots of the Terra satellite estimate of $PM_{2.5}$ from the generalized additive model and the chemical transport model versus the OpenAQ data, US Diplomatic Post in Delhi, India   124

C.4 Monthly average $PM_{2.5}$ for each method, Delhi, India   126

C.5 Annual average $PM_{2.5}$ for each method, by monitor site, Delhi, India   126

C.6 Residual evaluation plots for the generalized additive model trained on the Aqua satellite data, Ulaanbaatar, Mongolia   127

C.7 Plot of the dependence of the ground-level $PM_{2.5}$ estimate from the generalized additive model on the ratio of the Aqua aerosol optical depth and the MERRA planetary-boundary-layer height, Ulaanbaatar, Mongolia   128

C.8 Scatterplots of the generalized-additive-model- and chemical-transport-model–based concentrations using the Aqua aerosol-optical-depth data, versus OpenAQ data, US Diplomatic Post in Ulaanbaatar, Mongolia   129

C.9 Monthly average $PM_{2.5}$ for each method, Ulaanbaatar, Mongolia   130

C.10 Annual average (March–November) $PM_{2.5}$ for each method by monitor site, Ulaanbaatar, Mongolia   130

C.11 Valid Terra aerosol-optical-depth retrievals for monitoring sites in the OpenAQ database, Lima, Peru, 2016–17   131

C.12 Scatterplot of the chemical-transport-model–based $PM_{2.5}$ concentrations using the Terra aerosol-optical-depth data, versus ground-level-monitoring $PM_{2.5}$ data, Malam Junction site in Accra, Ghana   132

C.13 Residual evaluation plots for the generalized additive model trained on the Aqua satellite data, Kathmandu, Nepal   133

C.14 Plot of the dependence of the ground-level $PM_{2.5}$ estimate from the generalized additive model on the ratio of the Aqua aerosol optical depth and the MERRA planetary-boundary-layer height, Kathmandu, Nepal   134

C.15 Scatterplots of the generalized-additive-model– and chemical-transport-model–based concentrations using the Aqua aerosol-optical-depth data, versus OpenAQ data, US Diplomatic Post in Kathmandu, Nepal   134

C.16 Residual evaluation plots for the generalized additive model trained on the Terra satellite data, Addis Ababa, Ethiopia   135

C.17 Plot of the dependence of the ground-level $PM_{2.5}$ estimate from the generalized additive model on the ratio of the Terra aerosol optical depth and the MERRA planetary-boundary-layer height, Addis Ababa, Ethiopia   136

C.18 Scatterplots of the generalized-additive-model– and chemical-transport-model–based concentrations, using the Aqua aerosol-optical-depth data, versus OpenAQ Data, US Diplomatic Post in Addis Ababa, Ethiopia   137

C.19 Scatterplot of the chemical-transport-model–based $PM_{2.5}$ concentrations using the Terra aerosol-optical-depth data versus citywide average $PM_{2.5}$ data, Dakar, Senegal   137

C.20 Residual evaluation plots for the generalized additive model trained on the Aqua satellite data, Kampala, Uganda   139

C.21 Plot of the dependence of the ground-level $PM_{2.5}$ estimate from the generalized additive model on the ratio of the Aqua aerosol optical depth and the MERRA planetary-boundary-layer height, Kampala, Uganda   140

C.22 Scatterplots of the generalized-additive-model– and chemical-transport-model–based $PM_{2.5}$ concentrations, using the Aqua aerosol-optical-depth data, versus OpenAQ data, US Diplomatic Post, Kampala, Uganda   140

C.23 Residual evaluation plots for the generalized additive model trained on the Terra satellite data, Hanoi, Vietnam   141

C.24 Plot of the dependence of the ground-level $PM_{2.5}$ estimate from the generalized additive model on the ratio of the Terra aerosol optical depth and the MERRA planetary-boundary-layer height, Hanoi, Vietnam   142

C.25 Scatterplots of the generalized-additive-model– and chemical-transport-model–based concentrations using Terra aerosol-optical-depth data, versus OpenAQ data, US Diplomatic Post in Hanoi, Vietnam   143

C.26 Difference in base processing monthly average versus monthly averages after the quality assurance tool is applied   144

C.27 Generalized-additive-model residual evaluation plots from Terra quality assurance tool   145

C.28 Scatterplots of the statistical and chemical-transport-model–based concentrations using the Aqua aerosol-optical-depth data versus the QA filtered OpenAQ data, US Diplomatic Post in Ulaanbaatar, Mongolia   146

D.1 Picture of the Met One beta attenuation monitor unit showing the beta source and the filter tape used to collect the particulate matter samples   150

D.2 Schematic diagram of a tapered element oscillating microbalance   152

D.3 Tapered element oscillating microbalance–filter dynamics measurement system schematic   154

D.4 Time-series, seasonal, monthly, weekly, and diurnal $PM_{2.5}$ mass concentration data measured from the US Diplomatic Post in Delhi, India   157

D.5 $PM_{2.5}$ trends from data acquired at the Mandir Marg, New Delhi, India–DPCC monitoring site, 1.45 miles west of US Diplomatic Post   158

D.6 $PM_{2.5}$ trends from data acquired at the Shadipur, New Delhi, India–CPCB monitoring site, 4.75 miles northwest of US Diplomatic Post   158

D.7 $PM_{2.5}$ trends from data acquired at the Punjabi Bagh, Delhi, India–DPCC monitoring site, 6.91 miles northwest of US Diplomatic Post   159

D.8 $PM_{2.5}$ trends from data acquired at the NSIT Dwarka, Delhi, India–CPCB monitoring site, 11 miles southwest of US Diplomatic Post   159

D.9   PM$_{2.5}$ trends from data acquired at the Ramakrishna Puram, Delhi, India–DPCC monitoring site, 5.69 miles south-southwest of US Diplomatic Post   160

D.10   PM$_{2.5}$ trends from data acquired at the Anand Vihar, Delhi, India–DPCC monitoring site, 4.6 miles east-northeast of US Diplomatic Post   160

D.11   Time-series, seasonal, monthly, weekly, and diurnal PM$_{2.5}$ mass concentration data measured from the US Diplomatic Post in Lima, Peru   162

D.12   PM$_{2.5}$ trends from data acquired at the San Borja, Peru, monitoring site, 2.7 miles southwest of US Diplomatic Post   162

D.13   PM$_{2.5}$ trends from data acquired at the Campo de Marte, Peru, monitoring site, 5.4 miles northwest of US Diplomatic Post   163

D.14   PM$_{2.5}$ trends from data acquired at the Santa Anita, Peru, monitoring site, 3.9 miles north of US Diplomatic Post   163

D.15   PM$_{2.5}$ trends from data acquired at the Ate, Peru, monitoring site, 6.1 miles northeast of US Diplomatic Post   164

Maps

A.1   Fraction of good-quality attempted retrievals from Deep Blue and Dark Target algorithms showing differences in coverage over desert regions, and showing differences in coverage due to the quality checks applied in each algorithm   61

A.2   Radiosonde launch locations for 00:00 UTC, December 8, 2017   73

A.3   Aircraft observation coverage from AMDAR and MDCRS, December 8, 2017   74

D.1   Proximity and distribution of government-run air-quality monitoring stations around the US Diplomatic Post in Delhi, India   156

D.2   Illustration of the proximity of the US Diplomatic Post in Lima, Peru, relative to a subset of the nearest government-run assurance quality monitoring stations for which OpenAQ PM$_{2.5}$ data were available   164

Tables

ES.1   Cities included in this analysis   xvi

3.1   Strengths and weaknesses of the different daily satellite aerosol-optical-depth products publicly available in near real-time   13

3.2   Strengths and weaknesses of different approaches to converting satellite aerosol-optical-depth data into estimates of ground-level PM$_{2.5}$ concentrations   17

4.1   PM$_{2.5}$ instrument manufacturers' stated uncertainties   24

5.1   Statistics for the Delhi, India, satellite ground-level PM$_{2.5}$ products tested in this work   30

5.2   Statistics for the Ulaanbaatar, Mongolia, satellite ground-level PM$_{2.5}$ products tested in this work   30

5.3   Annual average PM$_{2.5}$ surface concentrations for Delhi, India, 2017   31

5.4   Cities included in this work   35

5.5   Correlation coefficient, mean normalized bias, and mean normalized gross error for each city   36

5.6   Lessons learned and recommended next steps to improve monitoring of PM$_{2.5}$ in each city to better understand and reduce the health impacts of PM$_{2.5}$ 40

6.1   Summary of typical problems in using satellites in PM$_{2.5}$ monitoring, the consequences, and the recommended actions to further the use of satellite data in PM$_{2.5}$ monitoring to better understand and reduce the health impacts of PM$_{2.5}$ in low- and middle-income countries   47

6.2   Recommendations for the use of satellite observations to supplement ground-level monitoring data in low- and middle-income countries   48

A.1   Summary of operational aerosol products   57

A.2   Case study of satellite aerosol-optical-depth applicability for three representative low- and middle-income country urban areas   68

B.1   Three focus cities where OpenAQ aggregates real-time local government particulate matter data overlap with US Diplomatic Post measurements   98

B.2   Contents of the subsampled aerosol-optical-depth product files   99

B.3   Mapping between the parsed data file variable names and the aerosol-optical-depth products in the L2 files   100

B.4   Region of interest data fields in the SEVIRI netCDF output files   101

B.5   MERRA-2 data file collections and extracted variables for input into the chemical transport model and statistical approach scripts   103

B.6   Statistics for the Delhi, India, satellite ground-level $PM_{2.5}$ products tested in this work   108

B.7   Statistics for the Ulaanbaatar, Mongolia, satellite ground-level $PM_{2.5}$ products tested in this work   110

B.8   Annual average $PM_{2.5}$ surface concentrations for Delhi, India, 2017   111

C.1   Cities included in this study   122

C.2   Coordinates and number of ground-level monitoring sites, by city   122

C.3   Delhi, India, evaluation statistics for the generalized-additive-model– and chemical-transport-model–based methods   125

C.4   Evaluation statistics for the generalized-additive-model– and chemical-transport-model–based methods, Ulaanbaatar, Mongolia   128

C.5   Evaluation statistics for the chemical-transport-model–based method using Terra data, Accra, Ghana   132

C.6   Evaluation statistics for the generalized-additive-model– and chemical-transport-model–based methods, Kathmandu, Nepal   135

C.7   Evaluation statistics for the generalized-additive-model– and chemical-transport-model–based methods, Addis Ababa, Ethiopia   136

C.8   Evaluation statistics for the chemical-transport-model–based method, Dakar, Senegal   138

C.9   Evaluation statistics for the generalized-additive-model– and chemical-transport-model–based methods, Kampala, Uganda   141

C.10 Evaluation statistics for the generalized-additive-model– and chemical-transport-model–based methods, Hanoi, Vietnam   143

C.11 Average $PM_{2.5}$ concentrations for each monitor site in Delhi, India, 2016–17   144

C.12 Generalized-additive-model output statistics of quality assurance tool versus base processing, Delhi, India   145

C.13 Evaluation statistics (using the filtered OpenAQ data) for the statistical and chemical-transport-model–based methods, Delhi, India   146

C.14 Kriging and co-kriging statistics, Delhi, India   147

C.15 Kriging and co-kriging statistics, Ulaanbaatar, Mongolia   147

Preface

Outdoor air pollution creates a significant health burden across the world. Exposure to outdoor airborne pollutant particles, called fine particulate matter or $PM_{2.5}$, leads to an estimated 4.2 million deaths a year.[1] Most of those deaths occur in low- and middle-income countries (LMICs). Understanding the severity of air pollution is a fundamental step in reducing its health burden. Most LMICs, however, do not have any infrastructure for measuring air quality and monitoring air pollution. Budgetary constraints and lack of technical expertise pose additional challenges.

There is increasing interest in harnessing Earth-orbiting satellite technology to fill critical gaps in air-quality data in LMICs. Satellite technology has been used with success in air-quality measuring applications in high-income countries, which typically have well-established, operational ground-level monitoring networks.

This report investigates whether satellite observations could be used to improve the monitoring of outdoor $PM_{2.5}$ in LMICs and, if so, to identify pathways for those countries to incorporate satellite data into their daily, city-scale $PM_{2.5}$ monitoring. Specifically, the report evaluates the performance of satellite observations in predicting ground-level concentrations of $PM_{2.5}$ in nine cities in LMICs.

The report's findings indicate that satellite-derived estimates of surface $PM_{2.5}$ in LMICs are not reliable. The average errors in satellite-based estimates tend to be very large, and reliability was limited by cloudiness, snow cover, landscape features, satellite coverage, and topography.

Consequently, satellites cannot be considered as a replacement for properly run and maintained ground-level monitoring networks in LMICs. This report shows that ground-level monitoring and satellite data are best thought of as complements to each other. For the purpose of protecting human health, governments of LMICs, and institutions such as the World Bank, should prioritize establishing or strengthening ground-level monitoring networks to measure $PM_{2.5}$ air pollution and other harmful air pollutants.

The results of this report have been published in a special issue of the peer-reviewed journal *Atmospheric Environment*: "Emerging Strategies to Fill the Gaps in Ground-level Air Quality Data in Low- and Middle-Income Countries" (Alvarado and others 2019; Awe and Hagler 2020).

NOTE

1. https://www.who.int/health-topics/air-pollution#tab=tab_2.

REFERENCE

Alvarado, Matthew, J., A. E. McVey, J. D. Hegarty, E. S. Cross, C. A. Hasenkopf, R. Lynch, E. J. Kennelly, T. B. Onasch, Y. Awe, E. Sanchez-Triana, and G. Kleiman. 2019. "Evaluating the Use of Satellite Observations to Supplement Ground-Level Air Quality Data in Selected Cities in Low- and Middle-Income Countries." *Atmospheric Environment* 218: 117016. https://doi.org/10.1016/j.atmosenv.2019.117016.

Awe, Yewande and Gayle Hagler, ed. 2020. "Emerging Strategies to Fill the Gaps in Ground-level Air Quality Data in Low and Middle Income Countries." https://www.sciencedirect.com/journal/atmospheric-environment/special-issue/10TGZBV0GCB.

About This Work

The analytical work in this report builds on a growing body of evidence that the World Bank is building to inform more effective and efficient pollution management interventions and harness the transition to a circular economy in low- and middle-income countries (LMICs). This growing body of evidence focuses on strengthening the knowledge base that will prompt dedicated action to tackle the forms of pollution that cause the most significant health and social costs in LMICs. It also advances interdisciplinary approaches to assess the linkages between pollution management and the circular economy and the World Bank's dual goals of eradicating poverty and promoting shared prosperity. Recent contributions from this body of work include (1) the monetary valuation of the global cost of mortality and morbidity caused by exposure to ambient fine particulate matter air pollution, (2) the bolstering of the case for establishment and strengthening of ground-level air-quality monitoring networks in LMICs, (3) the development of a systematic framework to support analysis of health impacts from land-based pollution, and (4) economic and financial instruments to support the transition to a circular economy.

Acknowledgments

This report was prepared by a team led by Yewande Awe with the core team comprising Gary Kleiman and Ernesto Sánchez-Triana (all from the World Bank). The background report was prepared by Matthew Alvarado, Jennifer Hegarty, Edward Kennelly, Richard Lynch, and Amy McVey (Atmospheric and Environmental Research, Inc.); Eben Cross and Timothy Onasch (Aerodyne Research, Inc.); and Christa Hasenkopf (Airglow Labs, formerly of OpenAQ). Shane Ferdinandus (World Bank) provided administrative support.

The team would like to acknowledge, with thanks, the valuable advice and inputs of peer reviewers Martin Heger and Karin Shepardson. This report also benefited from comments provided by World Bank staff, including Marcelo Bortman, Hocine Chalal, Stephen Dorey, Momoe Kanada, and Fernando Loayza, as well as several members of the World Bank's Pollution Management Global Solutions Group and other participants, during two brown bag lunch events organized to share the findings at different stages of the analytical work. The contributions of Phil Dickerson (US Environmental Protection Agency AirNow program) and Joanne Green (Ricardo Energy & Environment, formerly of the Clean Air Institute) are also appreciated.

The team is grateful for ground-level monitoring data on air quality provided by Emmanuel Appoh and staff at the Environmental Protection Agency of the Ghana Ministry of Environment, Science, Technology and Innovation (EPA Ghana); and André Jacques M. Ngor Dioh and Aminata Mbow Diokhané at the Center for Air Quality Management of the Senegal Ministry of the Environment and Sustainable Development.

This report is a product of the Environment, Natural Resources and Blue Economy Global Practice (ENB GP) of the World Bank. This work was conducted under the supervision of Juergen Voegele (Vice President, Sustainable Development); Karin Kemper (Global Director, ENB GP); Richard Damania (Chief Economist, Sustainable Development); Iain Shuker (Practice Manager, East Africa, ENB GP); Julia Bucknall (Senior Adviser, Eastern and Southern Africa); Benoit Bosquet (Regional Director for Sustainable

Development, East Asia and Pacific); and Christian Albert Peter (Practice Manager, Global Platform Unit, ENB GP).

The financial support of the World Bank–administered Pollution Management and Environmental Health multidonor trust fund in the preparation of this report is gratefully acknowledged.

Executive Summary

Outdoor air pollution accounts for an estimated 4.2 million[1] deaths around the globe, caused by exposure to fine, inhalable particles with an aerodynamic diameter less than or equal to 2.5 microns ($PM_{2.5}$, also called fine particulate matter). Reducing $PM_{2.5}$ air pollution is thus crucial for improving public-health outcomes.

Measuring and reducing the health impacts of $PM_{2.5}$ is especially challenging in low- and middle-income countries (LMICs). This is because LMICs tend to have limited funds for air quality monitors and limited local expertise in air quality. These shortcomings can undermine the effectiveness of a country's policies to improve its air quality.

Earth-orbiting satellites measure the amount of aerosols in the atmosphere with a metric called aerosol optical depth (AOD), which is based on the amount of light scattered and absorbed by all the aerosols over a given location (from the Earth's surface to space). The global coverage of these satellites and their decades-long records can potentially complement the sparse ground-level monitoring networks in LMICs. However, the performance of the techniques used to convert the satellite observations of AOD into estimates of ground-level $PM_{2.5}$ concentrations, and thus $PM_{2.5}$ exposures, is not well established in LMICs, where ground-level monitoring (GLM) data on $PM_{2.5}$ may be infrequent or absent.

It can be difficult to accurately estimate surface $PM_{2.5}$ concentration based on satellite data. The satellite is directly measuring the reflected sunlight from the Earth, which is affected not only by the aerosols in the atmosphere, but also by the amount of sunlight reflected by the ground (the surface reflectance). Furthermore, because the reflected sunlight is affected by all the aerosols in the atmosphere, determining the ground-level $PM_{2.5}$ concentration requires making an estimate of how the AOD (from all levels of the atmosphere) is related to the ground-level $PM_{2.5}$ concentration (that is, just the $PM_{2.5}$ located at the surface). Thus, to calculate ground-level $PM_{2.5}$ concentrations and exposures using satellite data, assumptions must be made regarding surface reflectance and the relationship between the AOD and ground-level $PM_{2.5}$ concentration. Errors or uncertainties in these assumptions will lead to errors and uncertainties in the satellite-derived $PM_{2.5}$ concentrations and exposures.

TABLE ES.1 Cities included in this analysis

CITY	COUNTRY	LOCATION	INCOME GROUP
Accra	Ghana	Coastal, low altitude	Lower middle
Addis Ababa	Ethiopia	Inland, high altitude	Low
Dakar	Senegal	Coastal, low altitude	Lower middle
Delhi	India	Inland, low altitude	Lower middle
Hanoi	Vietnam	Inland, low altitude	Lower middle
Kampala	Uganda	Inland but near lake, high altitude	Low
Kathmandu	Nepal	Inland, high altitude	Lower middle
Lima	Peru	Coastal, low altitude	Upper middle
Ulaanbaatar	Mongolia	Inland, high altitude	Lower middle

Source: World Bank.
Note: Income groups correspond to World Bank Country Classifications by Income Level: 2021–22. Income classifications are affected by several factors and are subject to change over time.

The goal of this report is to investigate if satellite observations could be used to improve $PM_{2.5}$ monitoring in LMICs and, if so, to identify pathways for LMICs to incorporate satellite data into their daily, city-scale $PM_{2.5}$ monitoring. A review of the scientific literature was conducted to identify different approaches used to combine satellite observations with GLM of $PM_{2.5}$ and to understand their reported performance. In addition, several methods for using satellite observations from publicly available sources to predict observed ground-level $PM_{2.5}$ daily average concentrations (and thus exposures) were tested for use in cities in LMICs. These methods were then applied to nine cities within LMICs (table ES.1) to identify patterns in the satellite performance with respect to city altitude, location, and other variables.

LITERATURE REVIEW

Many groups have successfully used different approaches to estimate ground-level $PM_{2.5}$ concentrations, and thus exposures, using satellite observations. These groups have generally looked at a continental-to-global scale (rather than the city-specific scale of interest to this work) and have predicted monthly and annual average $PM_{2.5}$ concentrations and exposures (rather than the daily average estimates of interest in this work). However, the errors in these satellite estimates of $PM_{2.5}$ concentrations can be large (about 50 percent).

Although many satellites measure AOD, only a smaller number of satellites cover the entire globe each day and provide free data, shortly after the observation is made, that are available for use in LMICs for daily observations. In addition, many satellite data sets of AOD can be difficult to access, either because the user must pay for the data or because the download process cannot be easily automated. Because the estimated errors in the satellite AOD products are similar, the satellite product that offers the best balance of AOD coverage and fine horizontal resolution for a given city should be used. For this report, the National Aeronautics and Space Administration's (NASA) MODIS combined Deep Blue and Dark Target AOD data set was used, since it is free, available shortly after the observation is made, easy to access, has been extensively validated, and has good coverage over urban and coastal areas.

Previous studies have attempted to convert satellite AOD measurements to ground-level $PM_{2.5}$ estimates using statistical techniques, chemical transport model (CTM)–based approaches, or hybrid approaches.

- CTM-based approaches can be applied to any region of the globe and do not require any GLM data. Public global CTM data sets, such as the NASA MERRA-2 reanalysis, can be used for estimates of aerosol vertical profiles and other parameters. However, these CTM data sets can have significant errors in the CTM's simulation of how aerosol concentrations change with altitude.
- Statistical approaches do not depend on a CTM, but they require a multiyear, accurate GLM data record. Only some LMICs have sufficient GLM data to use these approaches.
- Hybrid approaches that correct CTM-based approaches with statistical models tend to give the best performance. However, they also require a multiyear, accurate GLM data record over many different land-use types (for example, urban and rural sites, coastal and inland sites), which few LMICs have.

Interpolation approaches, such as co-kriging and land-use regression, can be used to combine satellite $PM_{2.5}$ estimates with GLM data to provide estimates of $PM_{2.5}$ concentrations at a high horizontal resolution (100–500 meters) across urban areas. However, these approaches tend to work better as the satellite horizontal resolution is increased and as the GLM network covers a wider variety of sites. The highest horizontal resolution possible for current satellite AOD measurements is between 2 and 4 kilometers (for geostationary satellites) or 1 and 10 kilometers (for polar-orbiting satellites). Consequently, finer-scale predictions of ground-level $PM_{2.5}$ will be possible only for areas where extensive, high-quality GLM data exist.

PERFORMANCE OF SATELLITE APPROACHES IN LMICs

This report found that satellites cannot be a complete replacement for a GLM network. The CTM-based and statistical approaches for converting satellite AOD to ground-level $PM_{2.5}$ concentrations did a poor job of predicting the day-to-day or site-to-site variations in daily average $PM_{2.5}$ values within a city. In addition, the average errors in all satellite-based estimates of the daily average $PM_{2.5}$ concentration at a given location in a city tended to be very large (21–77 percent for the statistical methods, and 48–85 percent for the CTM-based methods), much larger than the 10 percent error that is usually considered acceptable for $PM_{2.5}$ monitoring (Alvarado and others 2019). These large errors indicate that satellite-derived estimates of $PM_{2.5}$ concentrations in LMICs are not reliable.

Many cities also had significant limitations in the availability of satellite observations of aerosols throughout the year. For example, no satellite observations are available in Ulaanbaatar, Mongolia, for the high-$PM_{2.5}$ winter months of December to mid-March because of persistent snow cover. Thus, even a perfect method for converting satellite AOD measurements to ground-level $PM_{2.5}$ estimates would underestimate the true annual average $PM_{2.5}$ concentration for Ulaanbaatar by a factor of two.

xviii | GETTING DOWN TO EARTH

The work reported here found some patterns in the performance of satellite methods in different cities. Satellite-based methods appear to work best for low-altitude, inland cities such as Delhi, India, and Hanoi, Vietnam, but still have significant errors (43–60 percent) in predictions of daily average $PM_{2.5}$ concentrations at sites within these cities. Coastal cities (including cities near large lakes) have poor satellite coverage, either because of persistent clouds or the mixture of land and water surfaces in the satellite measurement of AOD. CTM-based satellite approaches tend to underestimate $PM_{2.5}$ in high-altitude cities (except for Addis Ababa, Ethiopia, likely because of its location near the Sahara). Although this work found that, under some conditions, adding satellite data to GLM network data via co-kriging may reduce the number of GLM sites needed to characterize $PM_{2.5}$ concentrations within a city, this ability varies from city to city and could result in large errors in annual average estimates for cities with a seasonal or persistent lack of satellite AOD coverage.

The 2017 annual average $PM_{2.5}$ concentrations in Delhi and Ulaanbaatar derived with the CTM-based method used in this project were similar to the 2016 annual averages estimated in the uncorrected[2] Global Burden of Disease (GBD) 2016 data set.[3] However, at the city level, results are considerably less accurate. For example, in Ulaanbaatar the CTM-based method tested in this report and the uncorrected GBD 2016 data set each underestimates the true annual average $PM_{2.5}$ concentration by a factor of 10. This is likely because cities in appreciably different air quality environments than their surroundings, such as cities in mountain valleys surrounded by rural land like Ulaanbaatar, will not get accurate relationships between satellite AOD and ground-level $PM_{2.5}$ from the relatively coarse resolution of global CTMs.

The geographically weighted bias correction used in the GBD 2016 data set corrects for the fact that Ulaanbaatar is surrounded by a region that is much less polluted; but the corrected GBD 2016 data set still underestimates the true annual average by a factor of two. In addition, the correction requires the use of GLM data at a continental scale. If only the GLM data from a given city are used, the land-use parameters are not highly correlated with the $PM_{2.5}$ variations in the city. Estimating variations in annual average $PM_{2.5}$ within a city is therefore unlikely to be possible with satellite AOD data alone, especially for cities affected by the satellite-coverage issues noted above. Estimating fine-scale geographic variations of $PM_{2.5}$ in a city has only been done using extensive GLM network data (covering both urban and rural sites) and land-use regression, with the satellite-based $PM_{2.5}$ estimate used as a variable in the land-use regression.

RECOMMENDATIONS

Based on these results, GLM and satellite data are best thought of as complements to each other. Many GLM networks could be improved by considering satellite data, but all approaches using satellite data improve as the number of high-quality GLM sites is increased. Thus, it is important that LMICs strengthen support for the establishment of GLM networks to measure air pollutants that cause mortality, notably $PM_{2.5}$, in LMICs in Sub-Saharan Africa and other regions. These GLM networks must have adequate quality assurance and quality control and follow standard operating procedures to ensure the data are of sufficient quality to be used to estimate $PM_{2.5}$ exposures for health studies and to be combined with the satellite estimates.

In other words, satellite data may be useful for estimating average air quality for countries or large geographical areas. However, for the purpose of protecting human health, LMICs need to prioritize the establishment or strengthening of GLM networks where they are lacking or weak, to measure air quality at the level where human activity is typically carried out and people are exposed to air pollutants, notably $PM_{2.5}$, that are harmful to health and cause death.

This work also provides the following recommendations for the use of satellite observations to supplement GLM data in LMICs (see proposed LMIC typology in chapter 1):

- For countries with no GLM data (Type I), the only possible approach to convert AOD to ground-level $PM_{2.5}$ is a CTM-based approach. The satellite-derived ground-level $PM_{2.5}$ values should be assigned a high uncertainty that reflects not only the uncertainty in the AOD (about 20 percent) but also the estimated uncertainty in the CTM-derived AOD-to-$PM_{2.5}$ relationship (about 50 percent).

- For countries with a small amount of GLM data with variable quality (Type II), it is possible to derive a bias estimate for the CTM-based estimates of ground-level $PM_{2.5}$. However, it will not be possible to ascertain whether the bias-corrected estimate is truly more accurate than the raw CTM-based estimate, and thus it is advisable to report and store both values. This will allow for reprocessing of the satellite estimates when more-rigorous quality assurance procedures are developed.

- Countries currently establishing GLM networks (Type III) can explore both statistical and bias-corrected CTM-based approaches for converting AOD to $PM_{2.5}$. CTM-based approaches likely would be best for these countries, but statistical approaches should also be tested before making a final decision.

- Countries with reliable, comprehensive GLM networks (Types IV and V) will be able to use satellite observations to fill in the gaps of the existing GLM networks. The more extensive GLM networks in these countries will allow for more-accurate estimates of the geographical and seasonal variation in the AOD-to-$PM_{2.5}$ relationship, and thus purely statistical approaches may outperform CTM-based estimates.

In addition, the results of this work suggest the following recommendations for LMICs and the US State Department for reporting the metadata for their GLM measurements to ensure the data are sufficiently accurate to be usefully combined with satellite estimates of $PM_{2.5}$:

- $PM_{2.5}$ measurements need to be provided along with information on the instrument/technique type, estimates of measurement uncertainties, relevant metadata, and operational history. It is important that each $PM_{2.5}$ measurement not only identifies the type of instrument used for the measurement but also tracks the instrument's relevant history of use and calibration. This includes all "meta" data from instrument data files, including "housekeeping" files.

- Any recorded results from tests relating to standard operating procedures and quality-assurance protocols should be collected and disseminated along with the $PM_{2.5}$ measurements to establish an instrument track record over time, providing a transparent, data-centric measure of instrument reliability and performance.

- Currently, US embassies do not publish their full quality-assurance protocol metadata along with the $PM_{2.5}$ levels, thereby reducing the use of these measurements as well-established and controlled reference measurements. Thus, it is recommended that US embassies publish their full quality-assurance protocol and relevant metadata along with the $PM_{2.5}$ levels.

ADDED VALUE OF THIS WORK

The analytical added value of this report includes the following:

- A review and synthesis of the scientific literature on using satellites to estimate ground-level $PM_{2.5}$ concentrations, with a focus on the relevance of the approaches for city-scale, daily-average $PM_{2.5}$ monitoring in LMICs
- An analysis of the performance of different satellite AOD measurements and methods for converting AOD to ground-level $PM_{2.5}$ concentrations from a local, city-specific perspective, including a discussion of the patterns in the performance due to city location and
- Recommendations for practitioners and policy makers in LMICs on how to improve their estimates of $PM_{2.5}$ exposure through expansion of their GLM networks, adoption of rigorous quality-assurance procedures, and incorporation of satellite data.

NOTES

1. https://www.who.int/health-topics/air-pollution#tab=tab_2.
2. The uncorrected GBD 2016 data set is the data set derived by relating satellite AOD to ground-level $PM_{2.5}$ before correction for persistent errors (biases) using GLM data and geographically weighted regression.
3. Downloaded from http://fizz.phys.dal.ca/~atmos/martin/?page_id=140.

REFERENCES

Alvarado, M. J., A. E. McVey, J. D. Hegarty, E. S. Cross, C. A. Hasenkopf, R. Lynch, E. J. Kennelly, T. B. Onasch, Y. Awe, E. Sanchez-Triana, and G. Kleiman. 2019. "Evaluating the Use of Satellite Observations to Supplement Ground-Level Air Quality Data in Selected Cities in Low- and Middle-Income Countries." *Atmospheric Environment* 218: 117016. https://doi.org/10.1016/j.atmosenv.2019.117016.

GBD 2016 Risk Factors Collaborators. 2017. "Global, Regional, and National Comparative Risk Assessment of 84 Behavioural, Environmental and Occupational, and Metabolic Risks or Clusters of Risks, 1990–2016: A Systematic Analysis for the Global Burden of Disease Study 2016." *Lancet* 390: 1345–422.

Abbreviations

ABI	Advanced Baseline Imager
AERONET	Aerosol Robotic Network
AHI	Advanced Himawari Imager
AMDAR	Aircraft Meteorological Data Relay
AOD	aerosol optical depth
AOT	aerosol optical thickness
AQ	air quality
BAM	beta attenuation monitor
BAMM	beta attenuation mass monitor
CLASS	Comprehensive Large Array-Data Stewardship System (NOAA)
CTM	chemical transport model
DAAC	Distributed Active Archive Center
ECMWF	European Centre for Medium-Range Weather Forecasts
EDGAR	Emission Database for Global Atmospheric Research
EDM	environmental dust monitor
EPA	Environmental Protection Agency (US)
FEM	federal equivalent method
FRM	federal reference method
GBD	Global Burden of Disease
GEOS	Goddard Earth Observing System
GFS	Global Forecast System
GLM	ground-level monitoring
GOES	Geostationary Operational Environmental Satellite
GPS	global positioning system
GWR	geographically weighted regression
JMA	Japan Meteorology Agency
JPSS	Joint Polar Satellite System
LAADS	Level-1 and Atmospheric Archive and Distribution System (NASA)
LEO	low earth orbit
Lidar	light detection and ranging
LMICs	low- and middle-income countries
LUR	land-use regression

LUT	look-up table
MACC	Monitoring Atmospheric Chemistry and Climate
MAIAC	Multi-Angle Implementation of Atmospheric Correction
MERRA	Modern Era Retrospective-analysis for Research and Applications
MISR	Multi-angle Imaging Spectroradiometer
MNB	mean normalized bias
MNGE	mean normalized gross error
MODIS	Moderate-Resolution Imaging Spectroradiometer
MPL	micro-pulse lidar
MSG	Meteosat Second Generation
NASA	National Aeronautics and Space Administration
NCEP	National Centers for Environmental Prediction
NDVI	normalized difference vegetation index
NEMS	NOAA Environmental Modeling System
NGAC	NEMS GFS aerosol component
NIR	near-infrared
NOAA	National Oceanic and Atmospheric Administration
NRT	near real-time
NWP	numerical weather prediction
PBL	planetary boundary layer
PBLH	planetary boundary layer height
$PM_{2.5}/PM_{10}$	particulate matter with an aerodynamic diameter less than or equal to 2.5 microns/10 microns, respectively
QA	quality assurance
QC	quality control
RMSE	root-mean-square error
RT	radiative transfer
SEDAC	Socioeconomic Data and Applications Center
SEVIRI	Spinning Enhanced Visible and Infrared Imager
SMAOD	suspended matter, aerosol optical depth
SMAOL	SEVIRI-MSG Aerosol Over Land
SPARTAN	Surface Particulate Matter Network
TEOM	tapered element oscillating microbalance
TOA	top of atmosphere
UN	United Nations
USGS	United States Geological Survey
UV	ultraviolet
VIIRS	Visible Infrared Imaging Radiometer Suite
VIS	visible
WMO	World Meteorological Organization

Glossary

Aerosol(s). Tiny solid and liquid particles suspended in the air. Windblown dust, sea salts, volcanic ash, smoke from wildfires, and pollution from factories are all examples of aerosols.

Aerosol optical depth (AOD) or aerosol optical thickness (AOT). A measure of the extinction (absorption and scattering) of light by aerosols in a column of air from the Earth's surface up to space. Defined as the natural logarithm of the ratio of the incident light at the top of the atmosphere to the transmitted light at the surface, assuming the light travels straight down.

Bias or mean bias. The average (mean) difference between a measurement and the true value of the property being measured. An ideal measurement has a mean bias of zero. If this value is large, it suggests that there are consistent (nonrandom) errors in the measurement.

Chemical transport model (CTM). A computer model that simulates the transport and chemistry of gases and aerosols in the atmosphere and can aid in air-quality forecasting and planning.

Co-kriging. An extension of kriging that uses a secondary data set to provide additional information on the spatial distribution of the air-quality variable being interpolated.

Correlation coefficient (R). In statistics, the proportion of the variance in the dependent variable that is predictable from the independent variable(s). If two measurements of $PM_{2.5}$ are well correlated such that they predict similar spatial and temporal patterns, then the R value will be near 1. If the values are uncorrelated, the value will be near 0.

Data assimilation. A process for using observations of the atmosphere (weather and chemical composition) to improve computer-model representations of the atmosphere.

Doppler wind profiler (DWP). A weather instrument that uses radar to detect the wind speed and direction at various elevations above the ground.

Generalized additive model (GAM). A statistical model of a system that is an extension of multiple linear regression models that allows the unknown functional form for each of the predictor variables to be determined from the data.

Geographic information system (GIS). A system designed to capture, store, manipulate, analyze, manage, and present geographic data. Usually analyzed using specialized software packages such as ArcGIS.

Geographically weighted regression. A statistical model of a system that is an extension of multiple linear regression models that allows the coefficients to vary with location.

Geostationary satellite. A satellite for which the orbit remains stationary over a given point on the Earth's equator. This type of satellite can continuously view the area underneath it but provides no data on other regions of the globe.

Global bias. Bias relative to a large set of observations, such as in several monitoring networks covering a continent.

Kriging. A method of data interpolation (that is, determining the value of a function at a given point by computing a weighted average of the known values of the function in the neighborhood of the point) used to provide estimates of air quality at locations without monitors, based on the measurements of, and distances from, surrounding air-quality monitors.

Land-use regression (LUR). A multiple linear regression model that uses different land-use variables (for example, population density, percent of urban land cover, and traffic intensity) to predict the concentration of air pollutants at a high resolution across an area using a sparser measurement network.

Lidar. Light detection and ranging; a measurement technique that uses reflected laser light to determine atmospheric properties such as the height of clouds and aerosol layers.

Local bias. Bias relative to a small set of nearby observations, such as the monitoring network within a city.

Look-up table (LUT). An approach to speeding up computer models that precalculates the values of many functions and stores them in a table for later use.

Mean normalized bias. Also called mean fractional bias, this is the average (mean) of the normalized bias, which is the difference between a measurement and the true value of the property being measured, divided by the true value. An ideal measurement has a mean normalized bias of zero. If this value is large, it suggests that there are consistent (nonrandom) errors in the measurement.

Mean normalized gross error. Similar to mean normalized bias, but the absolute value of the normalized bias is used so that positive and negative errors do not cancel out.

Near real-time (NRT) data. Model output or observational data that are processed and distributed as quickly as possible after the time being modeled or the time of the measurement.

Noise. In the context of a satellite measurement, noise refers to the random errors present in the measurement. Noise can generally be reduced by averaging over a larger area at the expense of a coarser horizontal resolution.

Particulate matter. See Aerosol(s).

Planetary boundary layer height (PBLH). The height of the well-mixed layer of air near the Earth's surface.

Polar-orbiting satellite. A satellite that has an orbit passing over or near the poles and thus can provide observations of the entire globe with the trade-off that each location is observed only (at most) twice a day.

Root-mean-square error (RMSE). An estimate of the average error in a measurement or model prediction that, unlike mean bias, does not allow positive and negative errors to cancel each other out.

Satellite AOD products. The estimate of the aerosol optical depth (AOD) over a given area produced from a given satellite instrument.

Surface reflectance. The fraction of sunlight that is reflected by the Earth's surface at a given location. Vegetation and ocean water tend to have low surface reflectance, while snow, deserts, and many urban surfaces have high surface reflectance.

1 Contextual Background and Objectives

This report was produced as part of a program of analytical work conducted under the framework of the World Bank's multidonor-funded Pollution Management and Environmental Health program. The objectives of this report are to improve knowledge regarding the following:

- How satellite measurements can best be used to enhance air-quality (AQ) monitoring, and thus improve human exposure assessment, in low- and middle-income countries (LMICs) and
- How satellite measurements can be brought into closer agreement with ground-level monitoring data, considering the shortcomings and advantages of satellite and ground-level measurements.

This report summarizes the findings of the three tasks: (1) a literature review of approaches used to combine satellite observations with ground-level monitoring measurements of particulate matter with an aerodynamic diameter less than or equal to 2.5 microns ($PM_{2.5}$), (2) the effort to develop and evaluate methods for converting satellite aerosol optical depth readings into ground-level $PM_{2.5}$ estimates in individual cities (Accra, Ghana; Delhi, India; Lima, Peru; and Ulaanbaatar, Mongolia), and (3) evaluation of the application of these methods to nine LMIC cities to identify geographic patterns in the performance of the satellite estimates.

Based on this work, recommendations are made for LMICs regarding how best to incorporate satellite data into their monitoring plans using the following proposed typology based on the country's level of engagement in AQ monitoring:

- *Type I:* Countries with no existing measurements and no history of any kind of routine measurements of atmospheric composition. Some anecdotal measurements or one-time sampling may have taken place.
- *Type II:* Countries with some information on atmospheric composition available (perhaps PM_{10}—particulate matter with an aerodynamic diameter of less than or equal to 10 microns—or total suspended particulates) but of variable quality without rigorous quality assurance procedures.

- *Type III:* Countries that possess reliable information but with poor spatial or temporal coverage; for example, monitoring may exist in only one city or routine monitoring may exist for a year or two but is no longer being collected because of equipment malfunction and/or lack of repairs.
- *Type IV:* Good, reliable AQ monitoring underway or being established.
- *Type V:* Routine, long-term AQ monitoring.

2 Introduction

Ambient or outdoor air pollution has significant impacts on public health around the globe. According to the World Health Organization, an estimated 4.2 million[1] people die every year worldwide from exposure to tiny solid and liquid particles—referred to as aerosols or fine particulate matter with an aerodynamic diameter less than or equal to 2.5 microns ($PM_{2.5}$)—suspended in outdoor air. Reducing these impacts through air pollution controls is thus a major goal of environmental efforts to improve public health. However, quantifying and reducing the health impacts of $PM_{2.5}$ is especially challenging in low- and middle-income countries (LMICs) because these countries tend to have limited air-quality (AQ) monitoring infrastructure, have insufficient quality assurance and quality control of their AQ monitoring data, and have limited local technical expertise in AQ monitoring, modeling, and planning. These shortcomings related to data and expertise can undermine the effectiveness of the design and implementation of policies to improve AQ in LMICs.

Earth-orbiting satellites can detect aerosols using observations of the sunlight scattered by the aerosols (figure 2.1; Bernard and others 2011; Levy and others 2013). This report examines whether the extensive spatial and temporal coverage of these satellites (figure 2.2) can potentially complement the sparse networks of ground-level monitors that typically exist in LMICs, thereby improving AQ monitoring and enforcement actions in these countries.

However, satellites are not able to measure ambient ground-level concentrations of $PM_{2.5}$ directly. Instead, satellites use the observations of the sunlight scattered by aerosols to estimate a parameter called aerosol optical depth (AOD), which represents the extinction (absorption and scattering) of light by all of the aerosols in a column of air from the Earth's surface up to the top of the atmosphere (for example, Levy and others 2013). As shown in figure 2.3, the reflected sunlight measured by the satellite depends on the intensity of sunlight reaching the Earth (I_o), the fraction of the sunlight reflected by the ground (α, also called the "surface reflectance" or "albedo"), the AOD, and the aerosol light scattering and absorbing properties (also called aerosol optical properties, here represented by a single factor ω). Although the intensity of the incoming sunlight reaching Earth is well known, the other three properties are more uncertain.

FIGURE 2.1

MODIS Aqua image of wildfire smoke being transported eastward to the Chesapeake Bay, US

Source: World Bank.
Note: The white haze image is produced by the scattering of sunlight by the aerosols in the smoke. Image taken June 10, 2015, 18:55 UTC. MODIS = Moderate-Resolution Imaging Spectroradiometer; UTC = Coordinated Universal Time.

To measure AOD, "retrieval algorithms" are designed that use tabulated estimates of the surface reflectance and aerosol optical properties.

Radiative-transfer (RT) models enable accurate simulations of reflectance at the top of the atmosphere (TOA) in the presence of aerosol layers with a variety of optical properties. By properly accounting for surface reflectance from land and ocean backgrounds, it is possible to retrieve several important aerosol properties by comparing satellite-observed TOA reflectances to those calculated from an RT model. Typically, to optimize processing time, TOA reflectances are precalculated by the RT model for several combinations of aerosols and are stored in look-up tables accessed by the retrieval algorithm. The RT calculations are performed for a range of aerosol optical thicknesses, so each stored reflectance value in the look-up table corresponds to an aerosol optical thickness.

The estimates of all three parameters (surface reflectance, AOD, and aerosol optical properties) are then refined to provide the best match with the observed reflected sunlight at multiple wavelengths (that is, the satellite measurement of the different colors in the reflected sunlight). However, uncertainties or errors in the estimated surface reflectance and aerosol optical properties can lead to similar uncertainties and errors in the measured AOD.

FIGURE 2.2

Current and future coverage provided by available polar and geostationary weather satellites

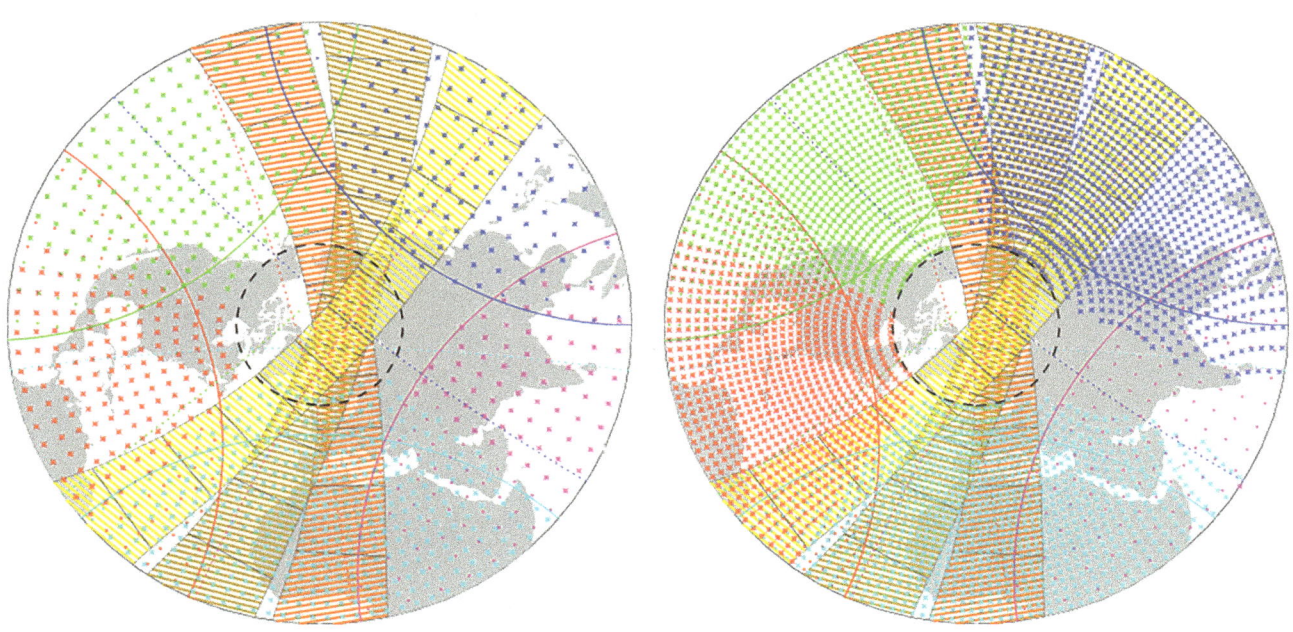

a. Current coverage provided by available polar and geostationary satellites

b. Future coverage after planned updates of geostationary satellites

Source: World Bank.
Note: The three bands in each panel illustrate three consecutive swaths from Moderate-Resolution Imaging Spectroradiometer (MODIS) in a polar orbit. Green, red, turquoise, magenta, and blue points represent coverage from geostationary satellites.

The exact relationship between AOD and ground-level $PM_{2.5}$ concentrations depends on the vertical distribution of aerosols above the observed location and their ability to scatter and absorb sunlight, which in turn depends on their size, shape, and chemical composition (van Donkelaar, Martin, and Park 2006). Two main approaches are used to estimate the relationship between AOD and ground-level $PM_{2.5}$ concentrations: statistical approaches (for example, generalized additive models [GAMs]) and chemical transport model (CTM)–based approaches. Statistical techniques can be used to predict the relationship between AOD and ground-level $PM_{2.5}$ using historical data sets for both parameters and other variables (for example, Hu and others 2014; Sorek-Hamer and others 2013, 2015; Strawa and others 2013). Alternatively, the relationship can be estimated by using a computer model of the chemistry of the atmosphere, called a chemical transport model (CTM) (for example, Geng and others 2015; van Donkelaar and others 2006, 2010, 2011, 2015a, 2015b).

Several factors can limit the ability of satellites to obtain accurate AOD measurements. Satellite observations of AOD generally require that the fraction of the sunlight reflected by the ground (that is, the surface reflectance) is relatively uniform across the satellite "footprint." This "footprint" is the area covered by a single satellite AOD observation, which may include multiple satellite image picture elements or "pixels." The horizontal extent of this footprint is referred to as the horizontal resolution of the AOD observation. Thus, mixed and reflective surfaces—including deserts, persistent snow cover, and mixed land and water

FIGURE 2.3

How satellites can measure aerosol optical depth from reflected sunlight

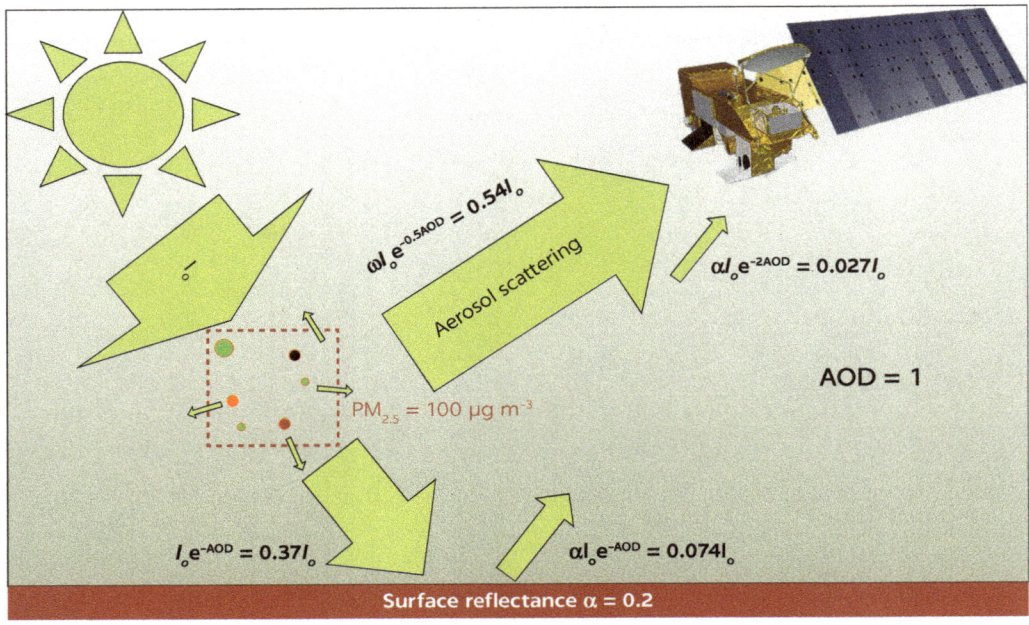

Source: World Bank.
Note: The reflected sunlight measured by the satellite includes that reflected by the aerosols (54 percent) and that reflected by the surface (2.7 percent). Here I_o represents the initial intensity of the sunlight, the surface reflectance is represented by one factor (α), and the aerosol scattering and absorbing properties are represented by a single combined factor (ω). The numbers are just to illustrate relative magnitudes and are not results from a radiative transfer model simulation. AOD = aerosol optical depth; $PM_{2.5}$ = particulate matter with an aerodynamic diameter less than or equal to 2.5 microns; $\mu g/m^3$ = micrograms per cubic meter.

surfaces near coastal cities—can reduce the ability of satellites to provide accurate AOD observations for many cities. Clouds also prevent satellite AOD observations, and thus cities with seasonal persistent clouds will have reduced satellite observations during those periods.

The statistical and CTM-based approaches for relating AOD to ground-level concentrations of $PM_{2.5}$ also have important limitations. The vertical profile of the aerosols, and thus the relationship between AOD and ground-level $PM_{2.5}$, depends strongly on the height of the well-mixed layer of air near the Earth's surface called the planetary boundary layer (PBL). For example, a one-kilometer PBL height with a more concentrated aerosol layer near the Earth's surface (green line in figure 2.4) may have the same AOD as a more dilute aerosol layer with a two-kilometer PBL height (blue line in figure 2.4). In addition, if there is a concentrated aerosol layer above the PBL—as can happen when wildfire, dust, or pollution is transported a long distance from its source—the AOD may be due mainly to this concentrated aerosol layer. In this case, the AOD observation does not provide much useful information about the aerosol concentration at the Earth's surface (red line in figure 2.4).

Lack of satellite AOD products at night, or due to clouds, snow cover, and other problems, can limit the ability of satellites to be used to estimate exposure to $PM_{2.5}$ for health studies. Satellite AOD observations are obtainable only for one specific time during the day (for polar orbiting satellites) or during the cloud-free daylight times (for geostationary satellites). However, the daily

FIGURE 2.4

Highly idealized example of three different aerosol vertical profiles, each with different ground-level PM$_{2.5}$ concentrations, that result in the same aerosol-optical-depth value of one when measured by satellites

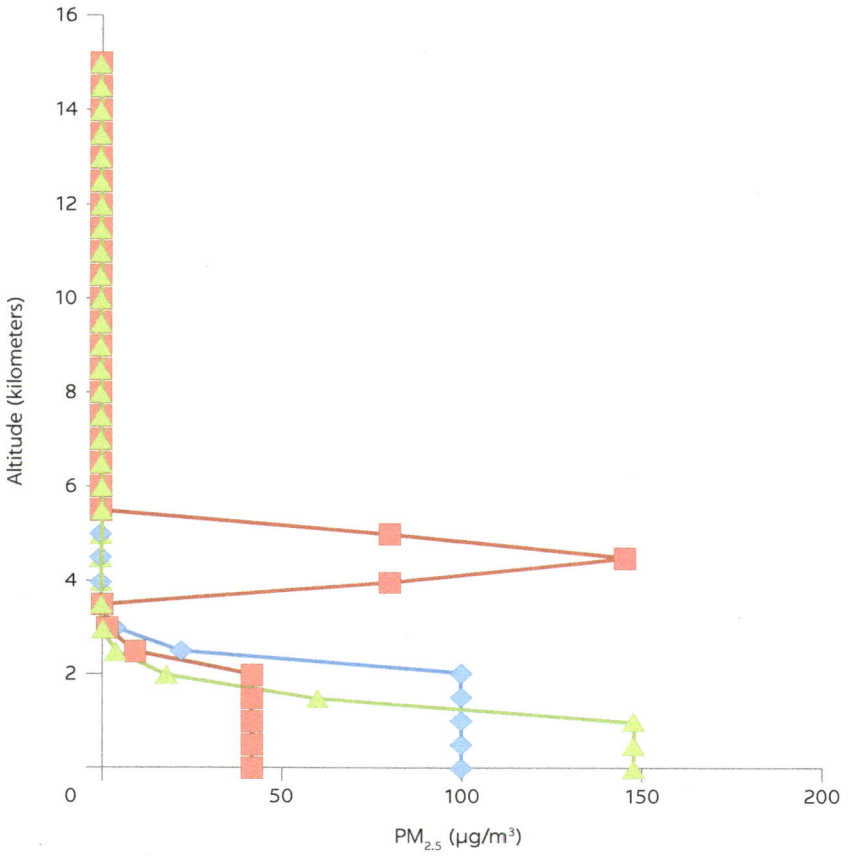

Source: World Bank.
Note: The green profile shows a 1-kilometer planetary boundary layer (PBL) height. The blue profile shows a 2-kilometer PBL height. The red profile shows a 2-kilometer PBL height and an elevated aerosol layer between 3.5 and 5.5 kilometers in altitude. AOD = aerosol optical depth; PM$_{2.5}$ = particulate matter with an aerodynamic diameter less than or equal to 2.5 microns; μg/m³ = micrograms per cubic meter.

average PM$_{2.5}$ concentration is the required metric for estimation of acute (short-term) health effects of PM$_{2.5}$. Consequently, even if the relationship between satellite AOD and the ground-level PM$_{2.5}$ *at the satellite observation time* were perfectly known, the relationship between the ground-level PM$_{2.5}$ concentration at the satellite observation time and the daily average PM$_{2.5}$ concentration would have to be estimated as well. This estimation would be based on either the statistics of past ground-level PM$_{2.5}$ observations or the output of a CTM. Errors in this relationship would lead to errors in the satellite-based estimate of PM$_{2.5}$ exposure.

In addition, investigations of the chronic (long-term) health effects of ground-level PM$_{2.5}$ require estimating the annual average PM$_{2.5}$ concentration. Thus, any seasonal pattern that affects the probability of a successful AOD retrieval, from either seasonal patterns in clouds or surface properties (that is, winter snow cover), will lead to a potentially incorrect annual average of

PM$_{2.5}$ from satellite observations. This is so even if the relationship between AOD and ground-level PM$_{2.5}$ concentrations is known.

The ability of satellite observations to estimate ground-level PM$_{2.5}$ has not been established in LMICs. In addition, most previous studies have provided global or continental-scale results on monthly or annual timescales, rather than the city-scale, daily average PM$_{2.5}$ concentration estimates needed for routine AQ monitoring and for public health alert programs. Most previous validation work on the use of satellite observations to estimate ground-level PM$_{2.5}$ has been performed in developed countries with extensive, well-calibrated, long-term ground-level monitoring (GLM) observations of PM$_{2.5}$ (for example, Geng and others 2015; Hu and others 2014; Sorek-Hamer and others 2013, 2015; Strawa and others 2013; van Donkelaar and others 2006, 2010, 2011, 2015a, 2015b). A few studies have evaluated the performance in some of the East Asia countries where GLM data are relatively new (for example, van Donkelaar, Martin, Brauer, and Boys 2015). However, the performance of the statistical models and CTMs used to derive these relationships has not been established in LMICs where GLM is infrequent or absent.

Thus, the satellite estimates of ground-level PM$_{2.5}$ in LMICs may be biased—that is, the average difference between the estimated ground-level PM$_{2.5}$ and the true PM$_{2.5}$ concentration may be large, suggesting consistent (nonrandom) errors in the satellite estimates. Techniques that attempt to remove these consistent errors are called bias-correction methods (for example, van Donkelaar, Martin, Spurr, and Burnett 2015), but these corrections depend on data from well-calibrated, long-term GLM networks and thus may not perform well in LMICs. Errors in the city-scale, daily average PM$_{2.5}$ concentration estimates could cause significant problems for public health alert programs, such as "code red" days when self-protective measures are advised for sensitive populations. This could result in significant economic impacts through lost work or school (if PM$_{2.5}$ is overestimated) or significant health, welfare, and productivity losses (if PM$_{2.5}$ is underestimated).

The lack of GLM data in many LMICs also introduces difficulties in trying to relate the spatially averaged AOD observations to variations in PM$_{2.5}$ within a city (due to distance from roadways and other pollution sources). This suggests the need to combine satellite observations with GLM observations and data on pollution sources within a city using interpolation techniques (for example, Lee, Chatfield, and Strawa 2016; Millar and others 2010; Vienneau and others 2013).

REPORT METHODOLOGY

This report evaluates different approaches for using satellite AOD data in estimating daily average ground-level PM$_{2.5}$ concentrations within selected cities in LMICs, with the goal of improving estimates of human PM$_{2.5}$ exposure. The methodological approach was, first, to conduct a literature review of previous approaches used to estimate PM$_{2.5}$ exposures employing satellite data, with a focus on identifying those satellite instruments and approaches for PM$_{2.5}$ estimation best suited to provide city-scale, daily average PM$_{2.5}$ concentrations and exposures. Nine LMIC cities with multiyear GLM data records were then identified to be used to test these approaches. Since the accuracy of these GLM data could affect the results of the analysis, the accuracy of the GLM instruments and the details of the quality-assurance procedures applied to these data were investigated. Finally, the identified satellite approaches (one statistical and one CTM-based) were applied to the nine cities to determine how the usefulness of satellite observations for estimating $PM_{2.5}$ concentrations and exposures varies across the geographic locations represented by the different cities.

OUTLINE OF THIS REPORT

This report discusses the results of a review of the literature on approaches used to combine satellite observations with GLM measurements of $PM_{2.5}$ and discusses the strengths and limitations of these approaches (chapter 3). The quality of GLM data available in LMICs is also discussed, and recommendations for improving the GLM data are made (chapter 4). In addition, this report discusses the effort to develop and evaluate methods for converting satellite AOD into daily average, city-scale, ground-level $PM_{2.5}$ estimates in nine cities in LMICs (chapter 5). Based on this work, recommendations are made for how LMICs could best incorporate satellite data into their ambient AQ monitoring efforts (chapter 6). The recommendations use the same typology proposed in chapter 1 based on a given LMIC's level of engagement with AQ monitoring. The appendixes provide additional information on the literature review conducted for this report (appendix A); converting satellite AOD to ground-level $PM_{2.5}$ (appendix B); evaluation of satellite approaches (appendix C); and quality considerations for GLM data in LMICs (appendix C).

NOTE

1. https://www.who.int/health-topics/air-pollution#tab=tab_2.

REFERENCES

Bernard, E., C. Moulin, D. Ramon, D. Jolivet, J. Riedi, and J.-M. Nicolas. 2011. "Description and Validation of an AOT Product over Land at the 0.6 μm Channel of the SEVIRI Sensor Onboard MSG." *Atmospheric Measurement Techniques* 4: 2543–65.

Geng, G., Q. Zhang, R. V. Martin, A. van Donkelaar, H. Huo, H. Che, J. Lin, and K. He. 2015. "Estimating Long-Term $PM_{2.5}$ Concentrations in China Using Satellite-Based Aerosol Optical Depth and a Chemical Transport Model." *Remote Sensing of Environment* 166: 262–70.

Hu, X., L. A. Waller, A. Lyapustin, Y. Wang, M. Z. Al-Hamdan, W. L. Crosson, M. G. Estes, S. M. Estes, D. A. Quattrochi, S. J. Puttaswamy, and Y. Liu. 2014. "Estimating Ground-Level $PM_{2.5}$ Concentrations in the Southeastern United States Using MAIAC AOD Retrievals and a Two-Stage Model." *Remote Sensing of Environment* 140: 220–32.

Lee, H. J., R. B. Chatfield, and A. W. Strawa. 2016. "Enhancing the Applicability of Satellite Remote Sensing for $PM_{2.5}$ Estimation Using MCDIS Deep Blue AOD and Land Use Regression in California, United States." *Environmental Science & Technology* 50 (12): 6546–55.

Levy, R. C., S. Mattoo, L. A. Munchak, L. A. Remer, A. M. Sayer, F. Patadia, and N. C. Hsu. 2013. "The Collection 6 MODIS Aerosol Products over Land and Ocean." *Atmospheric Measurement Techniques* 6: 2989–3034.

Millar, G., T. Abel, J. Allen, P. Barn, M. Noullett, J. Spagnol, and P. L. Jackson. 2010. "Evaluating Human Exposure to Fine Particulate Matter Part II: Modeling." *Geography Compass* 4 (7): 731–49.

Sorek-Hamer, M., I. Kloog, P. Koutrakis, A. W. Strawa, R. Chatfield, A. Cohen, W. L. Ridgway, and D. M. Broday. 2015. "Assessment of $PM_{2.5}$ Concentrations over Bright Surfaces Using MODIS Satellite Observations." *Remote Sensing of Environment* 163: 180–85.

Sorek-Hamer, M., A. W. Strawa, R. B. Chatfield, R. Esswein, A. Cohen, and D. M. Broday. 2013. "Improved Retrieval of $PM_{2.5}$ from Satellite Data Products Using Non-Linear Methods." *Environmental Pollution* 182: 417–23.

Strawa, A. W., R. B. Chatfield, M. Legg, B. Scarnato, and R. Esswein. 2013. "Improving Retrievals of Regional Fine Particulate Matter Concentrations from Moderate Resolution Imaging Spectroradiometer (MODIS) and Ozone Monitoring Instrument (OMI) Multisatellite Observations." *Journal of the Air & Waste Management Association* 63 (12): 1434–46.

van Donkelaar, A., R. V. Martin, M. Brauer, and B. L. Boys. 2015a. "Use of Satellite Observations for Long-Term Exposure Assessment of Global Concentrations of Fine Particulate Matter." *Environmental Health Perspectives* 123 (2): 135–43.

van Donkelaar, A., R. V. Martin, M. Brauer, R. Kahn, R. Levy, C. Verduzco, and P. J. Villeneuve. 2010. Global Estimates of Ambient Fine Particulate Matter Concentrations from Satellite-Based Aerosol Optical Depth: Development and Application." *Environmental Health Perspectives* 118 (6): 847–55.

van Donkelaar, A., R. V. Martin, R. C. Levy, A. M. da Silva, M. Krzyzanowski, N. E. Chubarova, E. Semutnikova, and A. J. Cohen. 2011. "Satellite-Based Estimates of Ground-Level Fine Particulate Matter during Extreme Events: A Case Study of the Moscow Fires in 2010." *Atmospheric Environment* 45 (34): 6225–32.

van Donkelaar, A., R. V. Martin, and R. J. Park. 2006. "Estimating Ground-Level $PM_{2.5}$ Using Aerosol Optical Depth Determined from Satellite Remote Sensing." *Journal of Geophysical Research: Atmospheres* 111: D21201.

van Donkelaar, A., R. V. Martin, R. J. Spurr, and R. T. Burnett. 2015b. "High-Resolution Satellite-Derived $PM_{2.5}$ from Optimal Estimation and Geographically Weighted Regression over North America." *Environmental Science & Technology* 49 (17): 10482–91.

Vienneau, D., K. de Hoogh, M. J. Bechle, R. Beelen, A. van Donkelaar, R. V. Martin, D. B. Millet, G. Hoek, and J. D. Marshall. 2013. "Western European Land Use Regression Incorporating Satellite- and Ground-Based Measurements of NO_2 and PM_{10}." *Environmental Science & Technology* 47 (23): 13555–64.

3 Literature Review of Approaches Used to Combine Satellite Observations with Ground-Level Monitoring in Urban Areas

The review of the scientific literature focused on identifying approaches to using satellite aerosol-optical-depth (AOD) observations that could provide city-scale, daily average $PM_{2.5}$ (particulate matter with an aerodynamic diameter less than or equal to 2.5 microns) monitoring in low- and middle-income countries (LMICs). This chapter discusses the approaches used in each of the three key steps in combining satellite observations with ground-level monitoring (GLM) of $PM_{2.5}$ for estimating ground-level concentrations in a given city:

1. The retrieval of AOD from satellite observations of reflected solar radiation (figure 2.3)

2. The translation of the vertically integrated AOD observation into an estimate of the ground-level $PM_{2.5}$ concentration, including any bias corrections (figure 2.4) and

3. Simultaneously interpolating these satellite-derived $PM_{2.5}$ estimates and the available GLM data to high-resolution grids centered on selected cities (figure 3.1).

The major findings of the review are presented below (see also table 3.1). The full literature review is provided in appendix A.

FIGURE 3.1

Annual average ground-level PM$_{10}$ concentrations, London, UK, and Rome, Italy

a. London

b. Rome

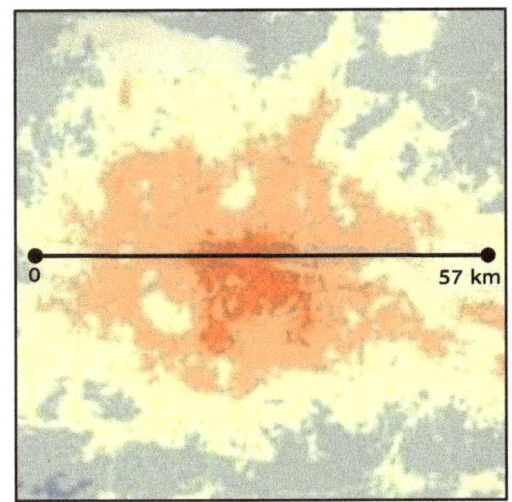
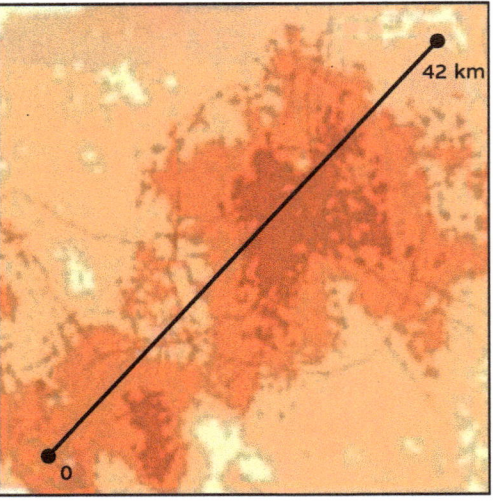

Source: Reproduced from Vienneau and others 2013.
Note: These images were created by combining ground-layer-monitoring data, satellite aerosol-optical-depth observations, and land-use data in a land-use regression model. km = kilometer; PM$_{10}$ = particulate matter with an aerodynamic diameter less than or equal to 10 microns.

RETRIEVAL OF AOD FROM SATELLITE OBSERVATIONS OF REFLECTED SOLAR RADIATION

Although many satellites provide AOD measurements using different approaches (the different estimates are referred to as "AOD products"), only a small number provide global coverage for each day free of charge shortly after the observation is made (that is, in near real-time [NRT]) at a horizontal resolution of 10 kilometers or finer. Table 3.1 summarizes the strengths and limitations of each of these identified NRT satellite AOD products with daily global coverage. The section "Available aerosol-optical-depth products" in appendix A presents a more comprehensive list of satellite products, including those not available in NRT and those that require subscription fees. For polar-orbiting satellites (that is, satellites that observe the entire globe, but at most twice a day), this includes the Moderate-Resolution Imaging Spectroradiometer (MODIS) Dark Target algorithm (10-kilometer and 3-kilometer resolution with daily global land and ocean coverage; Levy and others 2013; see figure 3.2); the MODIS Deep Blue algorithm (10-kilometer resolution with daily global land coverage; Hsu and others 2013); and the National Oceanic and Atmospheric Administration's Visible Infrared Imaging Radiometer Suite (VIIRS) AOD product (6-kilometer and 750-meter resolution with daily global land and ocean coverage; Jackson and others 2013). Only the MODIS products have had a comprehensive review of their performance, but comparisons with MODIS suggest the VIIRS 750-meter product tends to be biased high in urban and

TABLE 3.1 Strengths and weaknesses of the different daily satellite aerosol-optical-depth products publicly available in near real-time

AOD PRODUCT (AGENCY)	ORBIT	STRENGTHS	WEAKNESSES
MODIS Dark Target (NASA)	Polar	• Extensively validated • Global coverage • Easy to access • Includes AOD over both land and ocean	• Coarse horizontal resolution (10-kilometer—3-kilometer version much noisier) • Only two observations a day (10:30 and 13:30 local standard time) • No AOD over bright surfaces (deserts, cities)
MODIS Deep Blue (NASA)	Polar	• Extensively validated • Global coverage • Easy to access • Improved coverage over cities and deserts	• Coarse horizontal resolution (10-kilometer—3-kilometer version much noisier) • Only two observations a day (10:30 and 13:30 local standard time) • Only includes AOD over land
MODIS Combined (NASA)	Polar	• Extensively validated • Global coverage • Easy to access • Improved coverage over cities and deserts • Includes AOD over both land and ocean	• Coarse horizontal resolution (10-kilometer—3-kilometer version much noisier) • Only two observations a day (10:30 and 13:30 local standard time)
VIIRS (NOAA)	Polar	• Global coverage • Includes AOD over both land and ocean • Finest horizontal resolution for polar orbiting products (6-kilometer and 0.75-kilometer)	• Data difficult to access, download cannot be automated • 0.75-kilometer product very noisy and not yet extensively validated • Currently only one observation a day (13:30 local standard time) • No AOD over bright surfaces (deserts, cities)
MSG-SEVIRI SMAOL (EUMETSAT)	Geostationary	• Extensively validated • High resolution (3-kilometer) • Multiple observations a day • Easy to access	• Coverage only over Africa and Europe • AOD only over land; also excludes most coastal areas
GOES GASP (NOAA)	Geostationary	• Extensively validated • High resolution (4-kilometer) • Multiple observations a day	• Coverage only over North and South America • AOD only over land; also excludes most coastal areas • Data difficult to access, download cannot be automated
GOES SMAOD (NOAA)	Geostationary	• High resolution (2-kilometer) • Multiple observations a day	• Coverage only over North and South America • AOD only over land; also excludes most coastal areas • Data difficult to access, download cannot be automated • New product (starting 2018), needs further evaluation

Source: World Bank.
Note: In the context of a satellite measurement of AOD, *noise* refers to the random errors present in the measurement. Noise can generally be reduced by averaging over a larger area. AOD = aerosol optical depth; EUMETSAT = European Organisation for the Exploitation of Meteorological Satellites; GASP = GOES Aerosol/Smoke Product; GOES = Geostationary Operational Environmental Satellite; MODIS = Moderate-Resolution Imaging Spectroradiometer; MSG-SEVIRI = Meteosat Second Generation–Spinning Enhanced Visible and Infrared Imager; NASA = National Aeronautics and Space Administration; NOAA = National Oceanic and Atmospheric Administration; SMAOD = suspended matter, aerosol optical depth; SMAOL = SEVIRI-MSG Aerosol Over Land; VIIRS = Visible Infrared Imaging Radiometer Suite.

FIGURE 3.2

MODIS Dark Target 10-kilometer and 3-kilometer aerosol-optical-depth products retrieved for clear land and ocean fields of view and the local 5-kilometer average derived from the products, outer circle, compared to ground-based measurements, inner circle, over Baltimore, US

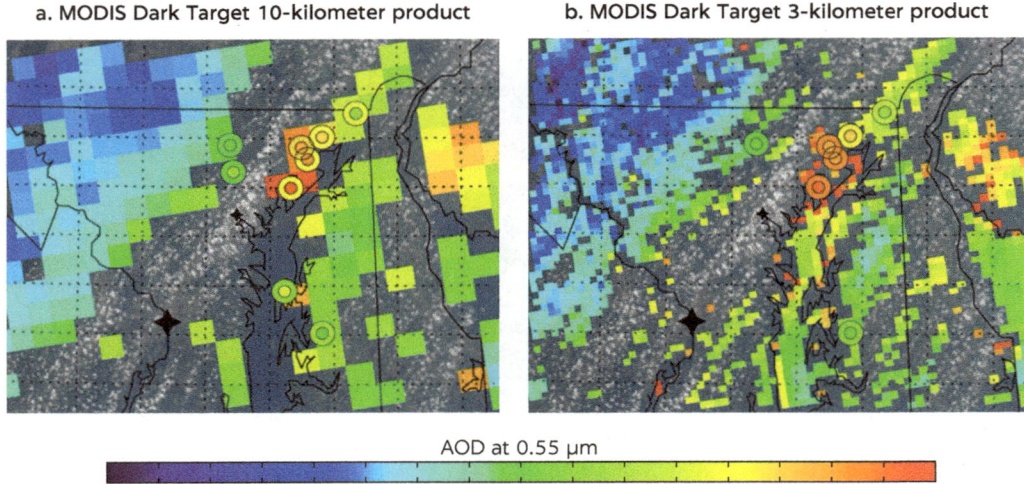

a. MODIS Dark Target 10-kilometer product

b. MODIS Dark Target 3-kilometer product

AOD at 0.55 μm

-0.05 0.05 0.15 0.25 0.35 0.45 0.55 0.65 0.75

Source: Munchak and others 2013. Note that these results used an older version of the MODIS surface reflectance scheme, and more recent versions (for example, Gupta and others 2016) show much better performance over urban areas.
Note: AOD = aerosol optical depth; MODIS = Moderate-Resolution Imaging Spectroradiometer.

mountainous regions. However, since the MODIS instruments are aging and new VIIRS instruments continue to be launched, the VIIRS products are more likely to be available in the coming decade. There are also several research products, such as extensions of the MODIS Dark Target and Deep Blue algorithms to VIIRS and the MODIS MAIAC algorithm (one-kilometer resolution with daily global coverage; Emili and others 2011; Lyapustin and others 2011) that may become operational in the near future. However, at the time this report was prepared, these products were available only for specific test time periods and locations.

For geostationary satellites (that is, satellites that continuously observe a fixed location; figure 3.3), the SMAOL (SEVIRI-MSG Aerosol Over Land) AOD product is available in NRT over Europe and Africa (three-kilometer resolution every 15 minutes; Bernard and others 2011; Mei and others 2012), and the GOES Aerosol/Smoke Product (GASP) AOD product is produced for the continental United States (four-kilometer resolution every hour; Knapp and others 2005). In 2018, the GASP product will be replaced with the GOES-16 SMAOD (suspended matter, aerosol optical depth) product[1,2] covering North and South America (two-kilometer resolution every 15 minutes). Himawari provides AOD retrievals over East Asia using an instrument such as GOES-16,[3,4] but the Himawari product is not freely available.[5,6]

Seasonal limitations in satellite coverage may result in biased estimates of annual average $PM_{2.5}$ concentrations from satellites. The potential

FIGURE 3.3

Example of good-quality aerosol optical depth 550 nm observations from a geostationary satellite from the US GOES-R SMAOD product

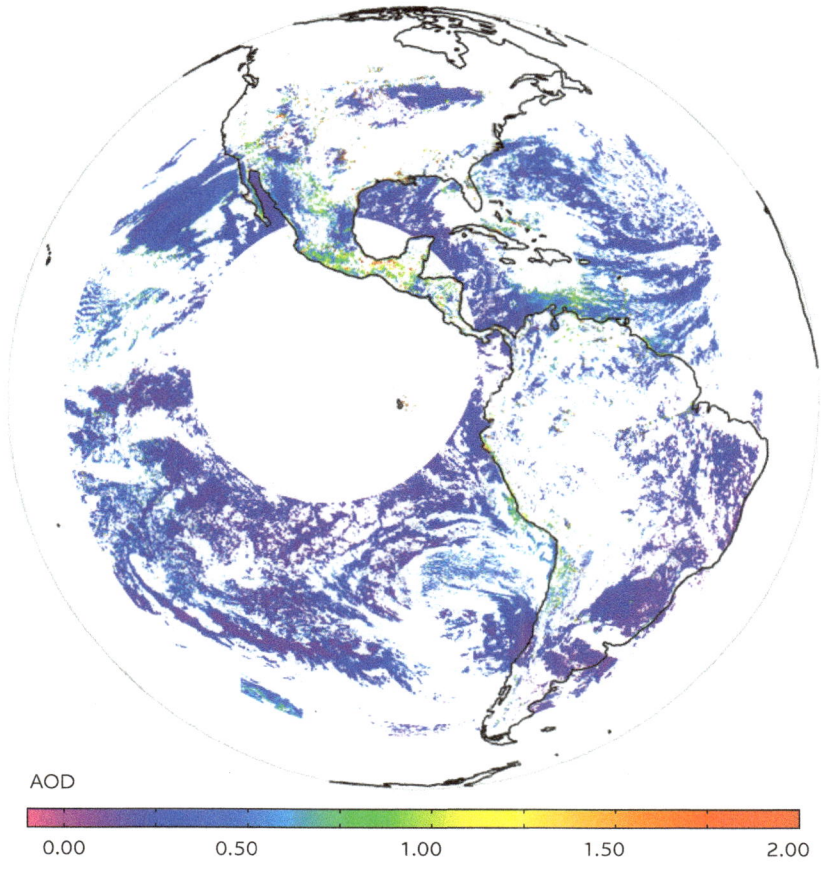

AOD

| 0.00 | 0.50 | 1.00 | 1.50 | 2.00 |

Source: World Bank.
Note: AOD = aerosol optical depth; GOES-R SMAOD = Geostationary Operational Environmental Satellite-R series–suspended matter, aerosol optical depth; nm = nanometer.

application of the currently available AOD products to three example cities (Delhi, India; Ulaanbaatar, Mongolia; and Lima, Peru) was examined as part of the literature review. None of these cities is currently covered by geostationary satellites with freely available AOD products, but the upcoming geostationary SMAOD product should cover Lima. In general, the MODIS Dark Target algorithm does not provide complete coverage over these cities because of highly reflective surfaces in urban areas, but the MODIS Deep Blue algorithm is able to provide AOD across the city. However, even with the Deep Blue algorithm, frequent clouds in Lima between May and November prevent AOD retrieval during much of this period (figure 3.4), and winter snow cover in Ulaanbaatar between December and March prevents AOD retrievals during this highly polluted season.

FIGURE 3.4

Valid MODIS Terra aerosol-optical-depth retrievals and Lima, Peru, monitoring sites in the OpenAQ database, 2016–17

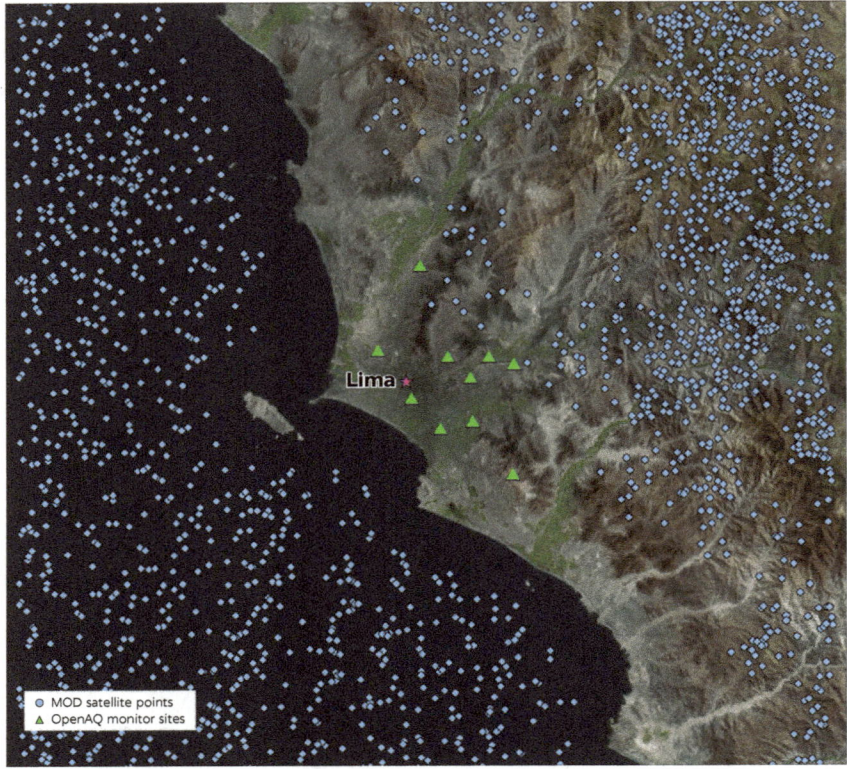

Source: World Bank, produced using Esri ArcGIS.
Note: The blue diamonds = valid satellite AOD data and the green triangles = OpenAQ monitoring sites. MOD = MODIS Terra; MODIS = Moderate-Resolution Imaging Spectroradiometer; OpenAQ = openaq.org.

THE TRANSLATION OF AOD TO GROUND-LEVEL PM$_{2.5}$ CONCENTRATIONS

Several studies have successfully converted satellite AOD to ground-level PM$_{2.5}$ estimates using statistical techniques, chemical transport model (CTM)–based approaches, or hybrid approaches, generally for continental-to-global spatial scales and monthly-to-annual time scales. Statistical approaches train statistical models, such as nonlinear generalized additive models (GAMs; Sorek-Hamer and others 2013; Strawa and others 2013), on historical GLM data to predict ground-level PM$_{2.5}$ using satellite AOD measurements and other meteorological and geographic data. CTM-based approaches use computer models of air quality, called "chemical transport models," to determine a time-varying relationship between ground-level PM$_{2.5}$ concentrations and satellite AOD measurements. This relationship is then used to scale the CTM aerosol profile until the CTM-calculated AOD matches the satellite AOD measurement, providing a better estimate of the ground-level PM$_{2.5}$ concentration than would be possible from the CTM alone. Hybrid methods combine statistical and CTM-based approaches by training a statistical model to correct the errors in the initial CTM-based satellite estimates of the ground-level PM$_{2.5}$ concentrations. For example, recent work has used a hybrid approach in which CTM-based approaches are followed by a second step that uses geographically weighted regression to correct for errors in the first-step estimates (van Donkelaar, Martin, Spurr, and Burnett 2015). These corrections require long-term (multiple years), reliable GLM data over a large region, including both urban and rural sites (see table 3.2).

Table 3.2 lists the strengths and weaknesses of these approaches. The statistical approaches (for example, Hu and others 2014; Sorek-Hamer and others 2013, 2015; Strawa and others 2013) can be more accurate if sufficient GLM data are available for the training. However, CTM-based approaches (for example, Geng and others 2015; van Donkelaar and others 2006, 2010, 2011; van Donkelaar, Martin, Brauer, and Boys 2015) are required in areas without GLM data or where available GLM data are of unreliable quality. Several potential sources of data are available for planetary boundary layer height and other meteorological parameters for the statistical approaches, as well as sources of CTM AOD and aerosol-profile data for the CTM-based approaches provided free of charge from global CTM run by high-income countries (see the section "Converting AOD to ground-level PM$_{2.5}$" in appendix A). LMICs, which often face competing needs for limited budgetary resources, may find that the use of freely available data is preferable to the expense of running their own CTMs or paying for data. In that case, the forecasts of ground-level concentrations and vertical profiles of PM$_{2.5}$ from the National Centers for Environmental Prediction Global Forecast System (Lu, da Silva, and others 2016; Lu, Wei, and others 2016) can be used if NRT output is needed, and the optimized historical ground-level concentrations and vertical profiles of PM$_{2.5}$ from the National Aeronautics and Space Administration MERRA-2 reanalysis (Provençal and others 2017) can be used for historical studies.

TABLE 3.2 **Strengths and weaknesses of different approaches to converting satellite aerosol-optical-depth data into estimates of ground-level PM$_{2.5}$ concentrations**

APPROACH	STRENGTHS	WEAKNESSES
CTM-based	• Easy to implement globally • Does not require any GLM data • Can remove cases with high-altitude aerosol plumes or low PBL heights from the analysis	• Can have significant biases due to errors in the CTM's simulation of the vertical profile and optical properties of the aerosols • Errors are likely larger for LMICs since estimates of PM$_{2.5}$ emissions are more out-of-date or uncertain for these countries • Errors are larger if horizontal resolution of the CTM is too coarse and thus a single model PM$_{2.5}$ value represents an average of PM$_{2.5}$ in urban and rural air
Statistical	• Locally unbiased (since these methods fit to the average for a single city) • Estimates error in the fit automatically • Does not require an initial guess at the aerosol profile or optical properties	• Requires a long-term (multiple years), reliable GLM data record with concurrent satellite observations • Cannot filter for high-altitude plumes • Determining appropriate statistical method for a given site can take many attempts
Hybrid	• Easy to implement globally • Globally unbiased (since these methods fit to the average for all cities and monitors considered) • Can filter the CTM results for high-altitude aerosol plumes or low PBL heights	• Requires a long-term (multiple years), reliable GLM data record with concurrent satellite observations • Requires GLM data in many different land-use types (urban and rural sites) • Can still have significant biases at specific sites

Source: World Bank.
Note: CTM = chemical transport model; GLM = ground-level monitoring; LMICs = low- and middle-income countries; PBL = planetary boundary layer; PM$_{2.5}$ = particulate matter with an aerodynamic diameter less than or equal to 2.5 microns.

COMBINING SATELLITE PM₂.₅ ESTIMATES AND GLM DATA INTO HIGH-RESOLUTION GRIDS CENTERED ON SELECTED CITIES

Co-kriging or land-use regression (LUR) can be used to combine satellite $PM_{2.5}$ estimates with GLM data to calculate estimates of $PM_{2.5}$ concentrations at a neighborhood scale (100 to 500 meters) across urban areas. Ordinary kriging uses a weighted average of neighboring GLM measurements to predict $PM_{2.5}$ concentrations across an urban area. Co-kriging is an extension of ordinary kriging that can take advantage of additional data sets or variables, using both the correlations between the monitors and the cross-correlations between the monitor data and the additional data sets (that is, the satellite-derived $PM_{2.5}$ estimates) to make better predictions (for example, Millar and others 2010; Pearce and others 2009; Wu, Winer, and Delfino 2006). LUR spatially links GLM measurements of $PM_{2.5}$ with other associated variables such as elevation, distance from roads, population density, and land-use type to develop fine-scale estimates of long-term $PM_{2.5}$ concentrations within an urban area that better represent gradients near highways and other pollution sources than is possible with kriging (for example, Lee, Chatfield, and Strawa 2016; Millar and others 2010; Vienneau and others 2013). LUR can also be used to predict the bias in CTM-based estimates of ground-level $PM_{2.5}$ concentrations as part of a hybrid approach to using satellites to estimate $PM_{2.5}$ exposures.

Co-kriging and LUR tend to work better as the satellite horizontal resolution is increased and as the GLM network covers a wider variety of sites (for example, urban and rural). Many GLM networks in LMICs are almost entirely in cities, which makes LUR more difficult because of the lack of variation in land-use parameters (for example, urban land fraction, distance of coasts, and distance from highways) across the network. Co-kriging could be used with satellite $PM_{2.5}$ estimates to reduce the number of GLM stations required to adequately cover an urban area (see discussion in chapter 5) and can be used to obtain either daily average or annual average estimates of $PM_{2.5}$. However, this study did not find any example in the literature where co-kriging had previously been used to combine satellite AOD and GLM $PM_{2.5}$ data. LUR has been successfully used by several groups to provide high horizontal resolution (100- to 500-meter), annual average estimates on $PM_{2.5}$ across regions and within cities using both satellite and GLM data (for example, Lee, Chatfield, and Strawa 2016; Vienneau and others 2013). Daily average estimates of $PM_{2.5}$ are also possible if meteorological data that vary daily and other data are used in the LUR. The needed data for LUR at 500-meter resolution is freely available for the entire globe, and most LUR studies have used the ArcGIS software package to perform the regression with geographical information system data sets.

NOTES

1. https://www.goes-r.gov/products/baseline-aerosol-opt-depth.html.
2. https://www.goes-r.gov/downloads/users/conferencesAndEvents/2014/GOES-R_Series_Program/04-Laszlo_abstract.pdf.
3. https://www.eorc.jaxa.jp/ptree/documents/Himawari_Monitor_Aerosol_Product_v5.pdf.
4. http://www.data.jma.go.jp/mscweb/technotes/msctechrep61-6.pdf.
5. http://www.jmbsc.or.jp/en/meteo-data.html.
6. http://www.jmbsc.or.jp/en/Data/Himawari-8-JMBSC-HP(2017.02.20).pdf.

REFERENCES

Bernard, E., C. Moulin, D. Ramon, D. Jolivet, J. Riedi, and J.-M. Nicolas. 2011. "Description and Validation of an AOT Product over Land at the 0.6 μm Channel of the SEVIRI Sensor Onboard MSG." *Atmospheric Measurement Techniques* 4: 2543–65.

Emili, E., A. Lyapustin, Y. Wang, C. Popp, S. Korkin, M. Zebisch, S. Wunderle, and M. Petitta. 2011. "High Spatial Resolution Aerosol Retrieval with MAIAC: Application to Mountain Regions." *Journal of Geophysical Research* 116: D23211.

Geng, G., Q. Zhang, R. V. Martin, A. van Donkelaar, H. Huo, H. Che, J. Lin, and K. He. 2015. "Estimating Long-Term $PM_{2.5}$ Concentrations in China Using Satellite-Based Aerosol Optical Depth and a Chemical Transport Model." *Remote Sensing of Environment* 166: 262–70.

Gupta, P., R. C. Levy, S. Mattoo, L. A. Remer, and L. A. Munchak. 2016. "A Surface Reflectance Scheme for Retrieving Aerosol Optical Depth over Urban Surfaces in MODIS Dark Target Retrieval Algorithm." *Atmospheric Measurement Techniques* 9: 3293–308.

Hsu, N. C., M. J. Jeong, C. Bettenhausen, A. M. Sayer, R. Hansell, C. S. Seftor, J. Huang, and S. C. Tsay. 2013. "Enhanced Deep Blue Aerosol Retrieval Algorithm: The Second Generation." *Journal of Geophysical Research: Atmospheres* 118 (16): 9296–315.

Hu, X., L. A. Waller, A. Lyapustin, Y. Wang, M. Z. Al-Hamdan, W. L. Crosson, M. G. Estes, S. M. Estes, D. A. Quattrochi, S. J. Puttaswamy, and Y. Liu. 2014. "Estimating Ground-Level $PM_{2.5}$ Concentrations in the Southeastern United States Using MAIAC AOD Retrievals and a Two-Stage Model." *Remote Sensing of Environment* 140: 220–32.

Jackson, J. M., H. Liu, I. Laszlo, S. Kondragunta, L. A. Remer, J. Huang, and H.-C. Huang. 2013. "Suomi-NPP VIIRS Aerosol Algorithms and Data Products." *Journal of Geophysical Research: Atmospheres* 118: 12673–89.

Knapp, K. R., R. Frouin, S. Kondragunta, and A. Prados. 2005. "Toward Aerosol Optical Depth Retrievals over Land from GOES Visible Radiances: Determining Surface Reflectance." *International Journal of Remote Sensing* 26 (18): 4097–116.

Lee, H. J., R. B. Chatfield, and A. W. Strawa. 2016. "Enhancing the Applicability of Satellite Remote Sensing for $PM_{2.5}$ Estimation Using MODIS Deep Blue AOD and Land Use Regression in California, United States." *Environmental Science & Technology* 50 (12): 6546–55.

Levy, R. C., S. Mattoo, L. A. Munchak, L. A. Remer, A. M. Sayer, F. Patadia, and N. C. Hsu. 2013. "The Collection 6 MODIS Aerosol Products over Land and Ocean." *Atmospheric Measurement Techniques* 6: 2989–3034.

Lu, C.-H., A. da Silva, J. Wang, S. Moorthi, M. Chin, P. Colarco, Y. Tang, P. S. Bhattacharjee, S.-P. Chen, H.-Y. Chuang, H.-M. H. Juang, J. McQueen, and M. Iredell. 2016. "The Implementation of NEMS GFS Aerosol Component (NGAC) Version 1.0 for Global Dust Forecasting at NOAA/NCEP." *Geoscientific Model Development* 9: 1905–19.

Lu, C.-H., S.-W. Wei, S. Kondragunta, Q. Zhao, J. McQueen, J. Wang, and P. Bhattacharjee. 2016. "NCEP Aerosol Data Assimilation Update: Improving NCEP Global Aerosol Forecasts Using JPSS-NPP VIIRS Aerosol Products." Paper presented at the "8th International Cooperative for Aerosol Prediction." http://icap.atmos.und.edu/ICAP8/Day3/Lu_NCEP_ThursdayAM.pdf.

Lyapustin, A., Y. Wang, I. Laszlo, R. Kahn, S. Korkin, L. Remer, R. Levy, and J. S. Reid. 2011. "Multiangle Implementation of Atmospheric Correction (MAIAC): 2. Aerosol Algorithm." *Journal of Geophysical Research* 116: D03211.

Mei, L., Y. Xue, G. de Leeuw, T. Holzer-Popp, J. Guang, Y. Li, L. Yang, H. Xu, X. Xu, C. Li, Y. Wang, C. Wu, T. Hou, X. He, J. Liu, J. Dong, and Z. Chen. 2012. "Retrieval of Aerosol Optical Depth over Land Based on a Time Series Technique Using MSG/SEVIRI Data." *Atmospheric Chemistry and Physics* 12: 9167–85.

Millar, G., T. Abel, J. Allen, P. Barn, M. Noullett, J. Spagnol, and P. L. Jackson. 2010. "Evaluating Human Exposure to Fine Particulate Matter Part II: Modeling." *Geography Compass* 4 (7): 731–49.

Munchak, L. A., R. C. Levy, S. Mattoo, L. A. Remer, B. N. Holben, J. S. Schafer, C. A. Hostetler, and R. A. Ferrare. 2013. "MODIS 3 km Aerosol Product: Applications over Land in an Urban/Suburban Region." *Atmospheric Measurement Techniques* 6: 1747–59.

Pearce, J. L., S. L. Rathbun, M. Aguilar-Villalobos, and L. P. Naeher. 2009. "Characterizing the Spatiotemporal Variability of $PM_{2.5}$ in Cusco, Peru Using Kriging with External Drift. *Atmospheric Environment* 43 (12): 2060–69.

Provençal, S., V. Buchard, A. M. da Silva, R. Leduc, N. Barrette, E. Elhacham, and S. H. Wang. 2017. "Evaluation of $PM_{2.5}$ Surface Concentrations Simulated by Version 1 of NASA's MERRA Aerosol Reanalysis over Israel and Taiwan." *Aerosol and Air Quality Research* 17: 253–61.

Sorek-Hamer, M., I. Kloog, P. Koutrakis, A. W. Strawa, R. Chatfield, A. Cohen, W. L. Ridgway, and D. M. Broday. 2015. "Assessment of $PM_{2.5}$ Concentrations over Bright Surfaces Using MODIS Satellite Observations." *Remote Sensing of Environment* 163: 180–85.

Sorek-Hamer, M., A. W. Strawa, R. B. Chatfield, R. Esswein, A. Cohen, and D. M. Broday. 2013. "Improved Retrieval of $PM_{2.5}$ from Satellite Data Products Using Non-Linear Methods." *Environmental Pollution* 182: 417–23.

Strawa, A. W., R. B. Chatfield, M. Legg, B. Scarnato, and R. Esswein. 2013. "Improving Retrievals of Regional Fine Particulate Matter Concentrations from Moderate Resolution Imaging Spectroradiometer (MODIS) and Ozone Monitoring Instrument (OMI) Multisatellite Observations." *Journal of the Air & Waste Management Association* 63 (12): 1434–46.

van Donkelaar, A., R. V. Martin, M. Brauer, and B. L. Boys. 2015. "Use of Satellite Observations for Long-Term Exposure Assessment of Global Concentrations of Fine Particulate Matter." *Environmental Health Perspectives* 123 (2): 135–43.

van Donkelaar, A., R. V. Martin, M. Brauer, R. Kahn, R. Levy, C. Verduzco, and P. J. Villeneuve. 2010. Global Estimates of Ambient Fine Particulate Matter Concentrations from Satellite-Based Aerosol Optical Depth: Development and Application." *Environmental Health Perspectives* 118 (6): 847–55.

van Donkelaar, A., R. V. Martin, R. C. Levy, A. M. da Silva, M. Krzyzanowski, N. E. Chubarova, E. Semutnikova, and A. J. Cohen. 2011. "Satellite-Based Estimates of Ground-Level Fine Particulate Matter during Extreme Events: A Case Study of the Moscow Fires in 2010." *Atmospheric Environment* 45 (34): 6225–32.

van Donkelaar, A., R. V. Martin, and R. J. Park. 2006. "Estimating Ground-Level $PM_{2.5}$ Using Aerosol Optical Depth Determined from Satellite Remote Sensing." *Journal of Geophysical Research: Atmospheres* 111: D21201.

van Donkelaar, A., R. V. Martin, R. J. Spurr, and R. T. Burnett. 2015. "High-Resolution Satellite-Derived $PM_{2.5}$ from Optimal Estimation and Geographically Weighted Regression over North America." *Environmental Science & Technology* 49 (17): 10482–91.

Vienneau, D., K. de Hoogh, M. J. Bechle, R. Beelen, A. van Donkelaar, R. V. Martin, D. B. Millet, G. Hoek, and J. D. Marshall. 2013. "Western European Land Use Regression Incorporating Satellite- and Ground-Based Measurements of NO_2 and PM_{10}." *Environmental Science & Technology* 47 (23): 13555–64.

Wu, J., A. M. Winer, and R. J. Delfino. 2006. "Exposure Assessment of Particulate Matter Air Pollution before, during, and after the 2003 Southern California Wildfires." *Atmospheric Environment* 40 (18): 3333–48.

4 Quality-Assurance Procedures in Low- and Middle-Income Countries

To better estimate the exposure of their citizens to particulate matter with an aerodynamic diameter less than or equal to 2.5 microns ($PM_{2.5}$) and the associated health effects, and to design policies to minimize these effects, low- and middle-income countries (LMICs) are working to begin, expand, and improve their air-quality monitoring efforts. However, some LMICs (Type I) currently have no air-quality data. Other LMICs (Type II) have air-quality data, but, because instrument maintenance and other "quality-assurance" procedures have not been fully followed, their data are potentially inaccurate. Still other LMICs (Type III) have accurate data but with insufficient spatial coverage. Satellites have been suggested as a way to reduce or eliminate the need for a ground-level-monitoring (GLM) network. However, accurate GLM data are needed to determine the performance of the satellite approaches to air-quality monitoring and to use the statistical and hybrid satellite approaches.

Thus, to evaluate the ability of satellites to assist with air-quality monitoring in LMICs (chapter 5), this chapter discusses the issues affecting the availability and accuracy of GLM data in LMICs in the first main section, the potential of automated methods to flag potentially incorrect GLM data in the second, and the need to examine the metadata, warning files, and operator logs for the GLM instruments to ensure the data are of sufficient accuracy to be used to develop and evaluate satellite approaches to $PM_{2.5}$ monitoring.

ISSUES AFFECTING THE AVAILABILITY AND ACCURACY OF GLM DATA IN LMICs

Recent high-air-pollution events in many LMICs have highlighted the need for improved dissemination of $PM_{2.5}$ concentrations obtained via GLM. Currently, one of the most promising dissemination mechanisms for real-time air-quality information in LMICs is the open-access web platform OpenAQ (see box 4.1). In the research underlying this report, we used data from OpenAQ for several LMICs to evaluate the ability of satellites to estimate daily average ground-level $PM_{2.5}$ concentrations in cities within LMICs. OpenAQ, and similar accessible

BOX 4.1

Open-access web platform for air quality data

OpenAQ (openaq.org) is an open-source platform and community that makes air-quality data accessible worldwide. The platform, launched in 2015, automatically aggregates, at 10-minute intervals, ground-level, near real-time air-quality data, shared publicly in disparate formats by government entities across the world. The data are synthesized into a universal format and made freely available via an application programming interface (API). To date, OpenAQ is the only site in the world to freely access station-level, physical particulate matter ($PM_{2.5}$, PM_{10}, BC) and gaseous (O_3, NO_2, SO_2, CO) air-quality data in aggregate, in near real-time or historically, that transparently traces back to their originating sources (for example, a government air-quality agency website, ftp server, or API). The open-source system is built and maintained by a community of software developers, scientists, open-data enthusiasts, and others across the world.

Note: BC = black carbon; CO = carbon monoxide; NO_2 = nitrogen dioxide; O_3 = ozone; $PM_{2.5}$/PM_{10} = particulate matter with an aerodynamic diameter less than or equal to 2.5 microns/10 microns, respectively; SO_2 = sulfur dioxide.

air-quality web platforms, can provide LMIC governments and citizens with important and timely access to local $PM_{2.5}$ concentrations.

However, simply rereporting $PM_{2.5}$ levels reported by LMICs does not guarantee that those data are of sufficient accuracy and precision for air-quality monitoring, or that those data can be usefully combined with satellite measurements to determine human exposures to $PM_{2.5}$. Many different instruments and techniques provide a measure of $PM_{2.5}$ with significantly different estimates of their accuracy (that is, how close they get to the true value) and precision (that is, how close two measurements of the same true concentration will be).

In addition, poor instrument operation, lack of calibrations, and incomplete quality assurance and quality controls can reduce the accuracy of $PM_{2.5}$ measurements. Given the seriousness of the health effects of $PM_{2.5}$, inaccurate measurements have the potential to mislead the public into a false sense of security (if the monitors are underreporting mass concentration) or incite alarm (if they are overreporting). These errors would also have economic impacts, such as reduced labor productivity due to people staying home during false alarms or getting ill by going out in missed poor air-quality conditions. Thus, it is important that GLM in LMICs provide reliable $PM_{2.5}$ measurements with well-understood uncertainties.

The instruments commonly used to measure $PM_{2.5}$ and report results to OpenAQ are approved by the US Environmental Protection Agency (EPA)[1] as federal equivalent methods (FEMs) for $PM_{2.5}$ characterization and thus are expected to provide daily average $PM_{2.5}$ concentrations within 10 percent of the true value when properly operated and calibrated. However, other "low-cost" instruments can have much higher errors. Common instruments for measuring $PM_{2.5}$ in use around world include (1) Met One Beta Attenuation Mass Monitor model BAM-1020; (2) Thermo Fisher Scientific model 5014i; (3) Thermo Fisher Scientific Tapered Element Oscillating Microbalance (TEOM, models 1400a, 1400b, and 1405); and (4) Grimm Aerosol Environmental Dust Monitor (EDM model 180). All these instruments have been approved by the EPA as FEMs for $PM_{2.5}$ characterization. To qualify as a "FEM," the $PM_{2.5}$ measurement technique must demonstrate differences of less than 10 percent relative to the federal reference method (FRM). In such head-to-head comparisons, the 10 percent threshold is applied to the intercept and slope of a correlation scatterplot of the test method versus the FRM. The EPA defines the FRM as gravitational (weight-based) $PM_{2.5}$ measurements, integrated or averaged over a 24-hour sample time. Figure 4.1 shows an example of the allowed uncertainties in the correlation (red hexagon outline) for the test instrument to qualify as a FEM. It is important to note the 24-hour averaging interval in the FEM designation. Given the dynamic and spatially heterogeneous nature of ambient $PM_{2.5}$ air pollution, it is often necessary to characterize $PM_{2.5}$ concentrations with faster time resolution (hourly or even one-minute data). Under these faster time-response conditions, the FEM status of a given $PM_{2.5}$ instrument may not hold.

Each of the common $PM_{2.5}$ measurement methods (BAMM, TEOM) rely on different techniques for measuring $PM_{2.5}$ (see appendix D) and thus have different precisions and accuracies. Although they output $PM_{2.5}$ mass concentrations, their measurement uncertainties vary for each device. Table 4.1 provides the manufacturer's stated uncertainties for each $PM_{2.5}$ instrument. Manufacturers provide precision uncertainties ($\mu g/m^3$) for a 1- and/or 24-hour average(s), stating that any single measurement over that averaging interval should be that close to the true value 99.7 percent of the time. Measurement uncertainties (for a well-operated system under the EPA requirements) would be the greater of these uncertainties or the 10 percent FEM error estimate for a given measurement. More details about the BAMM and TEOM methods are given in the sections "Operating methodologies and uncertainties associated with the BAMM technique" and "Operating methodologies and uncertainties associated with the TEOM technique," respectively, in appendix D.

Although it is important to know and understand the manufacturer's stated uncertainties, the true uncertainties of $PM_{2.5}$ measurements will be significantly larger, due to uncontrolled and unknown factors including site-specific

FIGURE 4.1

Example of US federal equivalent method measurement and uncertainties: Data set slope, intercept, and limits

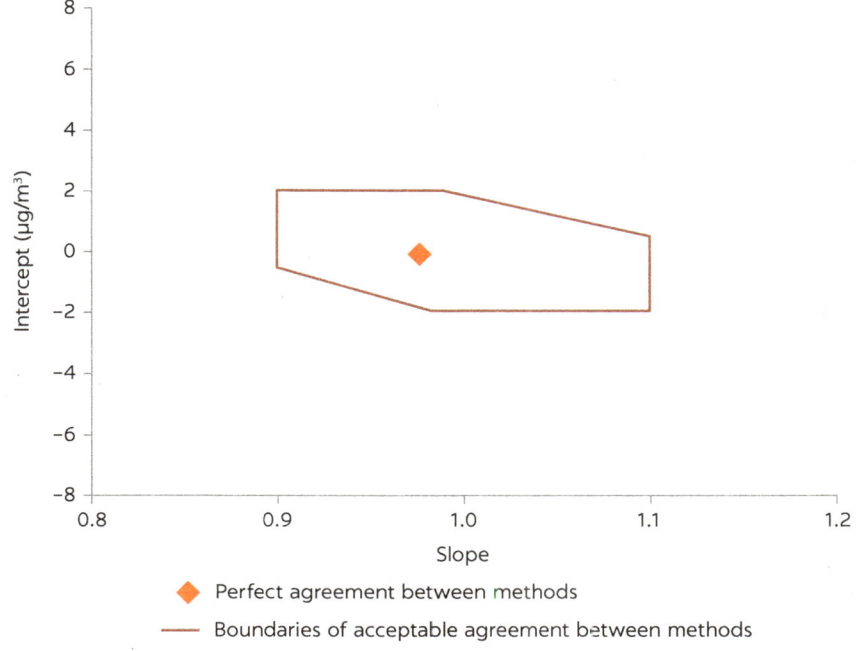

◆ Perfect agreement between methods

— Boundaries of acceptable agreement between methods

Source: Modified from US Code of Federal Regulations (Figure C-2 to Subpart C of Part 53, Title 40, 72 CFR 32204, June 12, 2007).
Note: $\mu g/m^3$ = micrograms per cubic meter.

TABLE 4.1 PM$_{2.5}$ instrument manufacturers' stated uncertainties

MANUFACTURER	INSTRUMENT/ MODEL	PRECISION (1 σ IN 1 HR)	UNITS	ACCURACY (1 σ IN 1 HR)	UNITS	COMMENTS
Grimm	EDM 180	—	—	—	—	Optical Particle Counter; provides only "reproducibility" value of ± 3%
MetOne	BAMS 1020	2.4	µg/m³	—	—	No accuracy provided
Thermo Fisher Scientific	TEOM 1400	2	µg/m³	0.75	%	None
Thermo Fisher Scientific	BAMS 5014i	2	µg/m³	5	%	None

Source: World Bank.
Note: PM$_{2.5}$ = particulate matter with an aerodynamic diameter less than or equal to 2.5 microns; µg/m³ = micrograms per cubic meter; — = not available; σ = standard deviation.

sampling considerations, volatile PM losses, operator errors, and others. These considerations can lead to inaccurate PM$_{2.5}$ data even when FEM instruments are used, and thus additional steps must be taken to ensure the accuracy of the GLM data. These steps can include the following:

- Automated processing of the PM$_{2.5}$ data to flag potentially erroneous data values (see the next section) and
- Examining instrument metadata, warning or error files, and operation logs to ensure that proper calibration, maintenance, and other quality-assurance procedures are being followed (see the section "Examining metadata, warning files, and operator logs" in chapter 4).

USING AUTOMATED SCRIPTS TO FLAG POTENTIALLY INCORRECT GLM DATA

To provide more information about the uncertainties and reliability of disseminated air-quality data (that is, in addition to manufacturers' stated precision uncertainty), OpenAQ created a data-quality wrapper to establish a baseline approach toward flagging potentially erroneous data values within the OpenAQ database (see also the section "Co-kriging and kriging with satellite and OpenAQ data" in appendix C). Guidance points for the wrapper development are outlined below:

(1) Develop flags on a per-measurement (rather than per-station) basis—Aim to justify "trustworthiness" on the basis of what is known or expected from a specific technique and avoid the negative political consequences of blacklisting specific sites altogether.
(2) Repeat data flags—When consecutive data points in the OpenAQ database from one PM$_{2.5}$ instrument repeat, generate a flag to examine the data more closely, since it is unlikely that identical data values are measured over time. Include the capacity for the user to input an adjustable parameter to define the number of repeat indexes necessary to trigger the flag.
(3) Zero points—If reported values are in fact 0.00000, generate a zero flag. Understand the extent to which the instrument is reporting zero due to an on-site zero calibration, instrument electronic glitch, or encountering PM$_{2.5}$ concentrations below the mass detection limits of the device.

(4) Negative values—Typically, instrument manufacturers and/or instrument operators will generate –999 or –9999 error flags within the output data stream to indicate a broader context on what was happening with the instrument at the time of the reported error code. This could correspond with on-site calibration (zero and span) or be an indicator that the system was off-line. Encourage OpenAQ users to flag and store these negative values within their initial QA/QC steps.

(5) Missing data—Within the OpenAQ database, one often finds periods with missing data (data not found at a given interval). The fact that the data are simply removed leaves some open questions, such as the following: How (if at all) do the missing data affect the resultant "average" values reported? What is the start and stop time for each reported average, and what fraction of this time was removed? In this context, "missing" data should be flagged, but not necessarily as –999. Tracking the extent of missing data for a given instrument may also be a useful indicator of that instrument's overall reliability.

EXAMINING METADATA, WARNING FILES, AND OPERATOR LOGS

To accurately evaluate satellite approaches and to combine satellite $PM_{2.5}$ estimates with GLM data, the GLM $PM_{2.5}$ measurements need to be provided in parallel with information on the instrument or technique type, estimates of measurement uncertainties, relevant metadata, and operational history. If the only publicly available information is the manufacturer and instrument type or model, one can identify rational lower-limit estimates of the expected uncertainties (assuming standard operational protocols for instrument upkeep are maintained). However, as noted above, the true uncertainties of $PM_{2.5}$ measurements can be significantly larger than the manufacturer's stated uncertainties. Without the necessary metadata to assess the quality of the $PM_{2.5}$ measurements, the statistical and hybrid approaches to satellite estimates of $PM_{2.5}$ cannot be used, and combining the GLM data with the satellite estimates may result in inaccurate estimates of $PM_{2.5}$ exposure and health effects.

Thus, it is important that each $PM_{2.5}$ measurement not only identifies the type of instrument used for the measurement, but also tracks the instrument's relevant history of use and calibration. This includes all "meta" data from instrument data files, including "housekeeping" files. Housekeeping files are often autogenerated by manufacturers to help identify deviation from normal operation. Such files would include error codes for alarm states, loss of sample flow, pressure changes, or periods where the instrument was out-of-spec (for example, temperature exceedances).

In addition, any recorded results from tests of standard operating procedures and quality-assurance protocols should be collected and disseminated along with the $PM_{2.5}$ measurements to establish an instrument track record over time providing a transparent, data-centric measure of instrument reliability and performance. This is typically done by following established protocols, such as defined by the EPA.

As determined during this project, US Diplomatic Posts are equipped to measure $PM_{2.5}$ with Met One beta attenuation monitor systems and follow the current EPA guidelines for $PM_{2.5}$ monitoring. However, their sampling location may not be representative of the typical air-quality conditions within a given city.

These EPA guidelines provide a well-established protocol for conducting $PM_{2.5}$ measurements with a high level of quality assurance and quality control. The US embassies typically provide these $PM_{2.5}$ data to host governments, and these data are now being collected on OpenAQ. Although it is important that these measurements get disseminated, it is important to stress that each measurement should be backed up with open access to the quality assurance and control (QA/QC) protocol metadata used to conduct the measurements and to QA/QC the measurements. Currently, US embassies do not publish their full QA/QC protocol metadata along with the $PM_{2.5}$ levels, thereby reducing the use of these measurements as well-established and controlled reference measurements. It is recommended that the US Diplomatic Posts publish their full QA/QC protocol and relevant metadata along with the $PM_{2.5}$ levels to make their data more useful.

NOTE

1. Other national air quality agencies do their own certification of methods. Most of the instruments listed are certified in multiple countries.

5 Using Satellite Data for Daily, City-Scale PM$_{2.5}$ Monitoring

As noted in chapter 2, many low- and middle-income countries (LMICs) do not have ground-level-monitoring (GLM) data of sufficient reliability for monitoring concentrations of particulate matter with an aerodynamic diameter less than or equal to 2.5 microns (PM$_{2.5}$). This leads to uncertainties related to the reliability of GLM data for use in designing policies to protect public health. It has been suggested that satellites could be used instead of GLM in these countries to monitor PM$_{2.5}$ or that satellites could be used in combination with GLM networks to improve the estimates of human exposure to PM$_{2.5}$. However, most previous studies of the potential of satellites to help with air-quality monitoring have been performed for countries with long-established, well-maintained GLM networks and have focused on global, annual average estimates of PM$_{2.5}$ concentrations and exposures (chapter 3). Thus, there is a need to provide guidance to LMICs on how satellite data can best be used in city-scale, daily air-quality monitoring; what the uncertainties (that is, expected errors) and biases of the satellite estimates of ground-level PM$_{2.5}$ concentrations are; and what local conditions (altitude, distance from coast, and so forth) lead to larger errors in the satellite estimates of ground-level PM$_{2.5}$ concentrations.

To address these needs, the results of the literature review (chapter 3) were used to identify promising new pathways for LMICs to use satellite observations in their city-scale, daily air-quality monitoring and forecasting. One of those novel approaches is statistical, and another is based on the chemical transport model (CTM); both approaches were tested. The statistical approach used the generalized additive model approach of Sorek-Hamer and others (2013), modified to use the ratio of the aerosol optical depth (AOD) to the planetary boundary layer (PBL) as the primary predictor of ground-level PM$_{2.5}$ concentrations, as in Chatfield and others (2017). To our knowledge, this particular statistical approach has not been tested before. The CTM-based approach used AODs and aerosol vertical profiles from the MERRA-2 reanalysis to derive a ground-level PM$_{2.5}$ estimate following the methods of van Donkelaar and others (2010). The use of standard CTM output freely provided by high-income countries for the CTM-based approach is novel, since most previous studies have invested significant effort in developing customized CTM simulations, and thus the performance of these approaches when using the model output available to LMICs has not been previously tested.

Subsequently, the GLM network data included in the OpenAQ database (chapter 4) was used to evaluate (1) the ability of satellites to predict daily average ground-level $PM_{2.5}$ concentrations and (2) their variations within cities, using these promising new approaches.

The next section describes the initial testing of the satellite approaches (using polar-orbiting satellite data) using 2017 GLM data sets for three cities: Delhi, India; Ulaanbaatar, Mongolia; and Lima, Peru. These cities were chosen because they have near real-time $PM_{2.5}$ observations at multiple locations in each city from local government air-quality agencies and US Diplomatic Post $PM_{2.5}$ data. A city in Africa (Accra, Ghana) was added to explore the ability of geostationary satellites to predict ground-level $PM_{2.5}$ concentrations as compared with polar-orbiting satellites. This work was performed to answer the following questions:

- Are the GLM data from the local government consistent with the data from the US Diplomatic Posts?
- How do the CTM-based and statistical-satellite estimates of ground-level $PM_{2.5}$ concentrations compare with the GLM data and with each other? Can either satellite approach correctly represent the day-to-day and site-to-site variation of $PM_{2.5}$ concentrations within a city?
- Do the satellite estimates of surface $PM_{2.5}$ improve when higher-resolution polar satellite or geostationary satellite data are used instead of the MODIS observations?
- Does a hybrid technique (using a statistical model to correct the biases in the CTM-based approach) perform better than a statistical or CTM-based approach alone?
- Can satellites reduce the number of required monitoring sites in a GLM network?

The last section describes the results of evaluating the satellite approaches for nine cities in LMICs using GLM data from 2016–17 where available. This work was performed to answer the following questions:

- How well do the satellite estimates of ground-level $PM_{2.5}$ concentrations compare with available GLM data in different cities?
- Can we identify patterns in the errors in the satellite estimates of ground-level $PM_{2.5}$ concentrations in the different cities based on their location (for example, distance from coasts, distance from deserts, and altitude)?

INITIAL TESTING OF SATELLITE PRODUCTS IN SELECTED CITIES

A full description of the methodology and results for this study is included in appendix B. The selected satellite data sets were from the MODIS (Moderate-Resolution Imaging Spectroradiometer) combined Deep Blue and Dark Target product (10-kilometer resolution, with data from the Aqua and Terra satellites tested separately), the standard VIIRS (Visible Infrared Imaging Radiometer Suite) AOD product (6-kilometer resolution), and the SEVIRI (Spinning Enhanced Visible and Infrared Imager) geostationary AOD product

(3-kilometer resolution, used as an example of the capability of next-generation geostationary satellites). Each satellite AOD data set also provides information about the quality of the AOD retrieval through a set of variables referred to as "quality flags" and provides recommendations for the minimum acceptable values for these flags. This is analogous to the quality filter for GLM data discussed in chapter 4. However, note that the quality flags only remove bad data; they do not provide information on the relative uncertainties in the remaining AOD measurements. All recommended quality flags were applied to filter the AOD data before analysis.

One statistical approach and one CTM-based approach were tested for each city for 2017 (except in Accra and Lima, where only a CTM approach could be used due to limited 2017 GLM data for these cities). Geographically weighted regression (GWR) was then performed to relate the errors between the GLM PM$_{2.5}$ observations and the two satellite-based estimates of ground-level PM$_{2.5}$ concentrations, as in van Donkelaar, Martin, Spurr, and Burnett (2015).

Consistency of local government and US Diplomatic Post data

In Delhi and Ulaanbaatar, the local GLM network data appear to be consistent with the US Diplomatic Post data, and thus the two data sets can be used together in the evaluation of the satellite methods. A test was performed in Delhi to determine if the US Diplomatic Post data were roughly consistent with the local GLM network data by using ordinary kriging of the local GLM network data to predict the annual average ground-level PM$_{2.5}$ concentration at the US Diplomatic Post. Kriging the local GLM network data captures the measured value at the US Diplomatic Post well, suggesting the two data sets are consistent. Further analysis (see the section "Interpreting ground-level PM$_{2.5}$ data from LMICs with limited metadata" in appendix D) suggested that the US Diplomatic Post data were consistent with local government monitors in Delhi but showed differences in Lima, where the local government monitors reported concentrations 30–50 percent lower than the US embassy data. Further investigation would be required to identify the sources of this difference.

Performance of different approaches for monitoring PM$_{2.5}$ with satellites

All cities had significant limitations in the availability of satellite AOD data that resulted in biased satellite estimates of annual average ground-level PM$_{2.5}$ concentrations. This is due to persistent clouds in Lima, wintertime snow cover in Ulaanbaatar, wet-season clouds in Delhi, and mixed water and land and bright surfaces in Accra. Satellite coverage was poorest in the coastal city of Lima and thus did not allow for a meaningful evaluation of a CTM-based approach for converting satellite AOD to ground-level PM$_{2.5}$ concentrations. In Ulaanbaatar, no satellite measurements are available for the high PM$_{2.5}$ winter months of December to mid-March, which would underestimate the true annual average PM$_{2.5}$ concentration for this city by 50 percent, even if satellites were able to perfectly predict ground-level PM$_{2.5}$ concentrations at other times. Delhi has satellite measurements in all months but substantially fewer observations in the peaks of the wet and dry seasons (December, January, July, and August), which would also result in a slight (about 10 percent) underestimate of the true annual average PM$_{2.5}$ concentration.

The CTM-based approach provided an accurate estimate (within about 10 percent) of the citywide annual average $PM_{2.5}$ concentration (necessary for assessing chronic health effect impacts) in Delhi, but the CTM-based approach underestimated the citywide annual average $PM_{2.5}$ concentration in Ulaanbaatar by a factor of 10. This is likely due to Ulaanbaatar's location in a river valley surrounded by mountains and mostly rural land. Thus, the coarse resolution of global CTMs (combined with likely inaccurate emission inventories for Ulaanbaatar) means that CTMs are not able to correctly represent the aerosol profile within the city, instead using an average profile that is more representative of the surrounding rural area. These same issues are present in the Global Burden of Disease (GBD) data set before the GWR correction with global GLM data is applied.[1] Statistical approaches were generally able to correctly represent the annual average $PM_{2.5}$ concentrations if the AOD data were available for the entire year.

The estimates of the ground-level daily average $PM_{2.5}$ concentrations from the CTM-based and statistical approaches generally had a low correlation ($R < 0.3$; tables 5.1 and 5.2) with the true daily average $PM_{2.5}$ concentrations within a city. Consequently, using satellite data alone would likely result in incorrect estimates of acute $PM_{2.5}$ exposure and health effects. The techniques also showed little efficacy in representing the variability in annual average $PM_{2.5}$ between surface sites, and so these techniques generally will not be able to identify areas of persistent high air pollution (that is, "hot spots") within a city.

Site-specific annual average values were not represented well in Delhi, limiting the usefulness of satellites to study variations in chronic health effects within the city (table 5.3). In addition, the lack of satellite AOD measurements

TABLE 5.1 Statistics for the Delhi, India, satellite ground-level PM$_{2.5}$ products tested in this work

	MODIS TERRA		MODIS AQUA		VIIRS S-NPP	
	STAT.	CTM	STAT.	CTM	STAT.	CTM
MB (micrograms per cubic meter)	0.045	−0.822	0.113	−29.275	−0.014	−58.102
MNB (%)	30.8	21.7	24.4	−7.5	32.3	−27.3
MNGE (%)	51.5	54.5	44.1	46.8	54.2	52.3
Correlation coefficient (R)	0.18	0.17	0.09	0.04	0.11	0.10

Source: Original calculations for this publication.
Note: CTM = chemical transport model–based method; MB = mean bias; MNB = mean normalized bias; MNGE = mean normalized gross error; MODIS = Moderate-Resolution Imaging Spectroradiometer; $PM_{2.5}$ = particulate matter with an aerodynamic diameter less than or equal to 2.5 microns; S-NPP = Suomi National Polar-Orbiting Partnership; STAT. = statistical method; VIIRS = Visible Infrared Imaging Radiometer Suite.

TABLE 5.2 Statistics for the Ulaanbaatar, Mongolia, satellite ground-level PM$_{2.5}$ products tested in this work

	MODIS TERRA		MODIS AQUA	
	STAT.	CTM	STAT.	CTM
MB (micrograms per cubic meter)	0.082	−31.174	0.443	−33.139
MNB (%)	54.0	−72.2	54.8	−73.1
MNGE (%)	78.5	78.2	78.3	78.8
Correlation coefficient (R)	0.17	0.00	0.30	0.01

Source: Original calculations for this publication.
Note: CTM = chemical transport model–based method; MB = mean bias; MNB = mean normalized bias; MNGE = mean normalized gross error; MODIS = Moderate-Resolution Imaging Spectroradiometer; $PM_{2.5}$ = particulate matter with an aerodynamic diameter less than or equal to 2.5 microns; STAT. = statistical method.

TABLE 5.3 **Annual average PM$_{2.5}$ surface concentrations for Delhi, India, 2017**
Micrograms per cubic meter

	GROUND AVERAGE	CTM-BASED AVERAGE	MERRA-2 AVERAGE
Anand Vihar	153.5	125.8	62.7
Delhi Technological University	123.9	113.5	58.5
Institute of Human Behavior and Allied Sciences	98.9	124.9	63.0
Income Tax Office	119.0	116.9	58.0
Mandir Marg	97.3	111.6	57.4
Netaji Subhas Institute of Technology Dwarka	140.2	112.9	62.4
Punjabi Bagh	103.6	117.0	55.8
Ramakrishna Puram	143.2	122.5	67.9
Shadipur	128.7	117.1	58.1
US Diplomatic Post	113.8	116.3	62.0
Urban average	**122.2**	**117.9**	**60.6**

Source: Original calculations for this publication.
Note: As determined by the ground-level-monitoring (GLM) data, the chemical transport model (CTM)–based satellite approach using MODIS (Moderate-Resolution Imaging Spectroradiometer) Terra aerosol optical depth (AOD), and the original MERRA-2 (Modern Era Retrospective-analysis for Research and Applications) output. Note these averages include only those days with both a valid AOD and a valid GLM daily average PM$_{2.5}$ (particulate matter with an aerodynamic diameter less than or equal to 2.5 microns) value.

in several months means that the satellites did a poor job of representing the observed seasonal cycle of PM$_{2.5}$ in Delhi.

Results of using alternative AOD products to estimate PM$_{2.5}$ concentrations

Using VIIRS instead of MODIS did not appreciably change the ability of the satellite approaches studied here to simulate the measured ground-level PM$_{2.5}$ concentrations in Delhi (figure 5.1). However, for cities where the morning MODIS Terra product performed better than the afternoon MODIS Aqua product, using the afternoon VIIRS product will also likely result in poorer predictions than using the lower-resolution MODIS Terra product, but this was tested directly only for Delhi.

The SEVIRI AOD product does provide a higher horizontal resolution than the MODIS product, but it appears to have lower coverage for coastal cities such as Accra, because of an AOD data quality filter that removes data near coastlines (to avoid mixed water and land surfaces; see figure 5.2) Thus, the SEVIRI AOD product is not suitable for use in coastal cities.

Using a hybrid technique to bias-correct CTM-based estimates

Land-use variables (percentage of urban land cover, elevation, and population density) had little ability to predict the bias in the CTM-based ground-level PM$_{2.5}$ estimates in Delhi or Ulaanbaatar ($R < 0.01$). This is likely because there are 10 or fewer GLM sites in each city within about 30 kilometers of each other, as opposed to the 1,440 sites used in the GWR bias correction of van Donkelaar, Martin, Spurr, and Burnett (2015) over the United States.

Further analysis suggests that the limitations of the satellite techniques in Delhi and Ulaanbaatar cannot be easily addressed using post hoc bias corrections. Applying a simple linear bias-correction model to the CTM-based PM$_{2.5}$

FIGURE 5.1

Comparison of the chemical-transport-model–based estimates for daily average ground-level PM$_{2.5}$ concentrations in micrograms per cubic meter over Delhi, India, using VIIRS and MODIS Terra, November 1, 2017

a. Using VIIRS

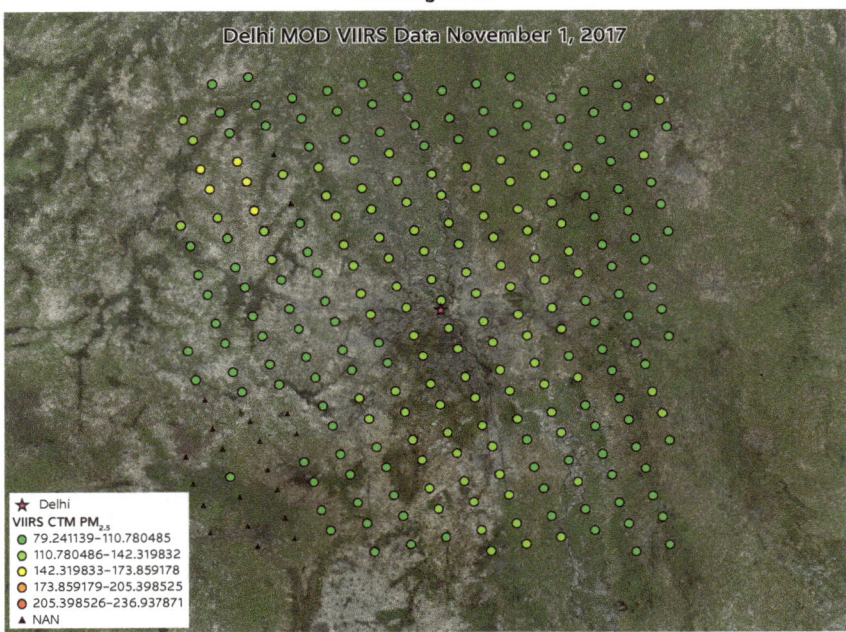

b. Using MODIS Terra

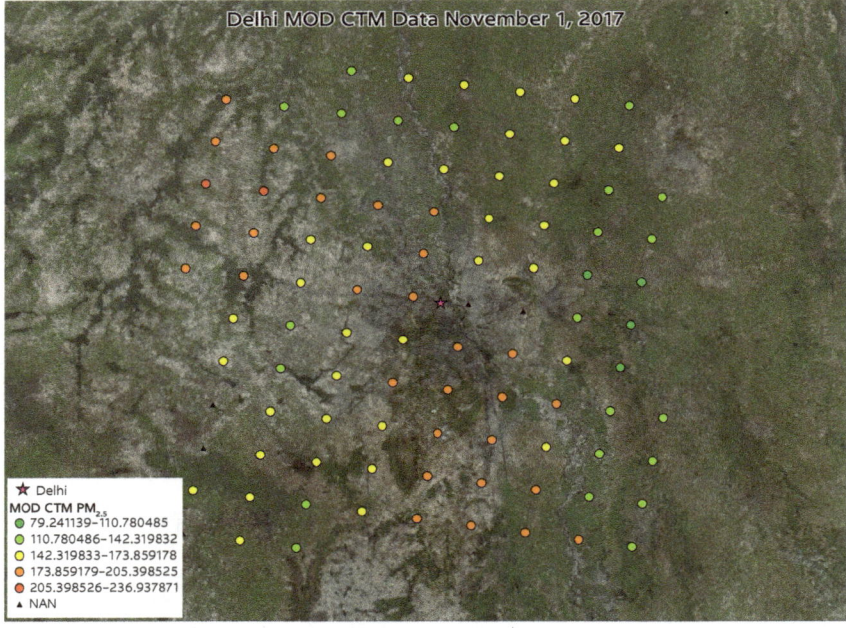

Source: World Bank, produced with Esri ArcGIS.
Note: CTM = chemical transport model-based method; MOD = MODIS Terra; MODIS = Moderate-Resolution Imaging Spectroradiometer; NAN = not a number; PM$_{2.5}$ = particulate matter with an aerodynamic diameter less than or equal to 2.5 microns; VIIRS = Visible Infrared Imaging Radiometer Suite.

FIGURE 5.2

Comparison of the chemical-transport-model–based estimates for daily average ground-level PM$_{2.5}$ concentrations in micrograms per cubic meter over Accra, Ghana, using SEVIRI and MODIS Aqua, December 26, 2017

a. Using SEVIRI

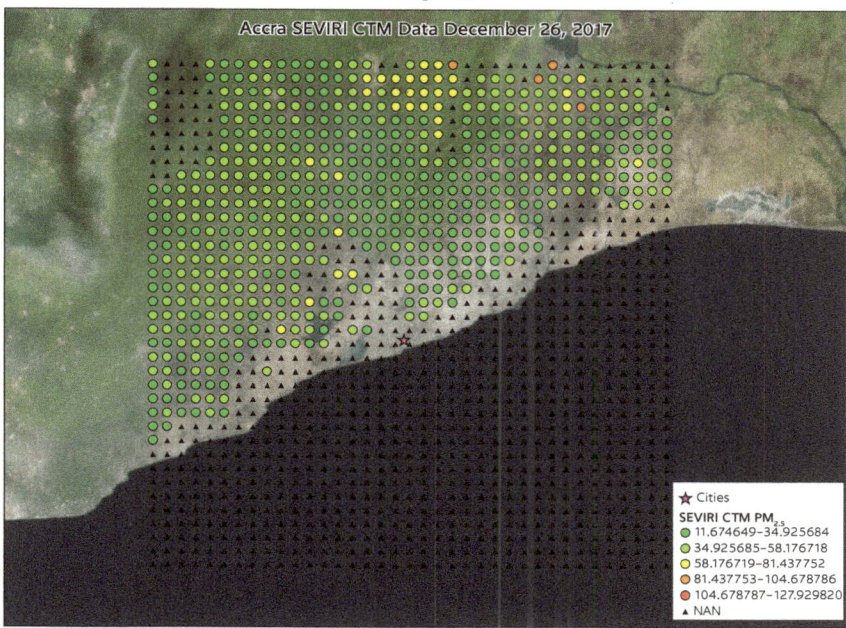

b. Using MODIS Aqua

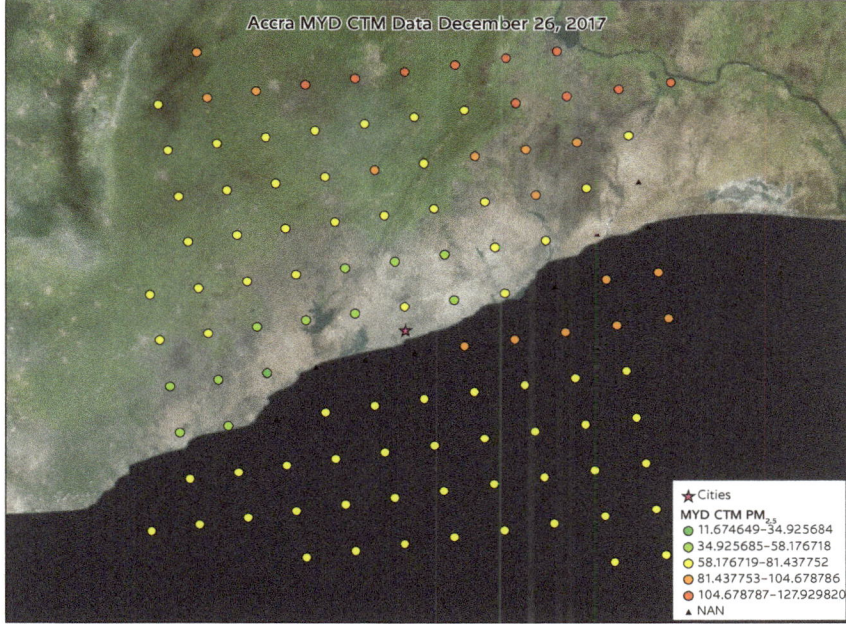

Source: World Bank, produced with Esri ArcGIS.
Note: CTM = chemical transport model-based method; MODIS = Moderate-Resolution Imaging Spectroradiometer; MYD = MODIS Aqua; NAN = not a number; PM$_{2.5}$ = particulate matter with an aerodynamic diameter less than or equal to 2.5 microns; SEVIRI = Spinring Enhanced Visible and Infrared Imager.

estimates did improve the mean bias in Delhi and Ulaanbaatar but did not greatly improve the ability of the land-use regression approach to predict the daily and site-to-site variation of $PM_{2.5}$ within an urban area. In Delhi, the Stage 1 CTM-based estimate was not a significant predictor of the daily average of $PM_{2.5}$ in the bias-correction model. In fact, a statistical model that included only the MERRA-2 speciated aerosol variable did as good a job at predicting the daily average of $PM_{2.5}$ in Delhi as a model using those MERRA-2 variables and the satellite Stage 1 CTM-based estimate.

Can satellites reduce the number of required monitoring sites in a GLM network?

Another goal of this work was to determine if satellite data could be combined with GLM data in such a way as to *reduce* the number of GLM sites required to quantify exposure in an urban population (see the section "Co-kriging and kriging with satellite and OpenAQ data" in appendix C). To test this, ordinary kriging was done for each day using the monitor station daily averages of $PM_{2.5}$ in Delhi and Ulaanbaatar excluding the US Diplomatic Post. The concentration at the US Diplomatic Post for each day was then predicted with ordinary kriging. This tested how well the other GLM sites alone could characterize the $PM_{2.5}$ concentrations at an unmeasured site. The CTM-based satellite $PM_{2.5}$ estimates were then combined with the GLM data (minus the US Diplomatic Post) via co-kriging, and we again estimated the concentrations at the US Diplomatic Post to test if the satellite data could compensate for the loss of one GLM station.

The Delhi results suggest that adding in the satellite data would not help to reduce the number of GLM monitoring sites there, since the root-mean-square error (a measure of prediction accuracy calculated as the square root of the average of the squared differences between the predicted and observed $PM_{2.5}$ concentrations at the US Diplomatic Post) increased above that for ordinary kriging when the CTM-based satellite data are added via co-kriging. This finding suggests that including the CTM-based satellite data resulted in poorer interpolation predictions than using GLM data alone. However, the Ulaanbaatar results suggest that the Diplomatic Post could be eliminated by adding satellite data to the rest of the GLM network in this city, but again only for the eight months each a year that satellites actually provide data. Overall, these results suggest that the use of satellite data may allow LMICs to reduce the number of GLM sites needed in an area, but that this ability varies from city to city and could result in large errors in annual average estimates for cities with a seasonal (Ulaanbaatar) or persistent (Lima) lack of satellite AOD coverage.

IDENTIFYING THE TYPE OF CITIES WHERE SATELLITES WORK BEST

We applied the CTM-based and statistical approaches discussed above to a set of nine cities in LMICs that represent a diversity of regimes (coastal versus inland, high versus low altitude, tropical versus temperate, and so forth), as shown in table 5.4. These cities also have GLM data needed for validation, through either US Diplomatic Posts, local monitoring networks, or both, although the amount of available GLM data varies significantly between cities.

TABLE 5.4 Cities included in this work

CITY	COUNTRY	LOCATION	INCOME GROUP	NUMBER OF GLM SITES
Accra	Ghana	Coastal, low altitude	Lower middle	4
Addis Ababa	Ethiopia	Inland, high altitude	Low	2[a]
Dakar	Senegal	Coastal, low altitude	Lower middle	4[b]
Delhi	India	Inland, low altitude	Lower middle	10
Hanoi	Vietnam	Inland, low altitude	Lower middle	1[a]
Kampala	Uganda	Inland but near lake, high altitude	Low	1[a]
Kathmandu	Nepal	Inland, high altitude	Lower middle	4
Lima	Peru	Coastal, low altitude	Upper middle	10
Ulaanbaatar	Mongolia	Inland, high altitude	Lower middle	8

Source: World Bank.
Note: GLM = ground-level monitoring.
a. Only US Diplomatic Post data are available.
b. Dakar data are available only as an average of the four reporting sites. Income groups correspond to World Bank Country Classifications by Income Level: 2021–22. Income classifications are affected by several factors and are subject to change in time.

For the nine cities, the satellite-based estimates of ground-level PM$_{2.5}$ concentrations used the MODIS Combined (Deep Blue plus Dark Target) Product,[2] since this provides the most coverage for cities, is easy to access, has been extensively validated, and has a long historical record. The tests above explored several other publicly available satellite sources of AOD but did not find any that provided significant advantages over this product. Both the CTM-based and statistical approaches discussed in the first main section of this chapter were applied to each city in table 5.4, and the mean bias and error in the satellite estimates for each city were calculated. The evaluation results are discussed in detail in appendix C and summarized below.

Evaluating satellite methods in multiple cities in LMICs

Table 5.5 and figures 5.3 to 5.5 show the statistics for the evaluations of the statistical and CTM-based methods for whichever satellite (MODIS Terra or MODIS Aqua) gave the best performance for a given city. The statistical methods generally performed better than the CTM-based estimates, but this may be due to the low numbers of points available for model training. The cities with the highest R also had the lowest number of evaluation points—for example, Kathmandu had an R of 66 percent, mainly because there were only 50 days in the two-year evaluation period (2016–17) that had both valid daily PM$_{2.5}$ averages from the GLM data and valid AOD retrievals from the satellite. This is an example of statistical model "overfitting" where, due to the small number of data points relative to the degrees of freedom of the model, the model corresponds too closely to the data with which the model was created. Consequently, the model may fail to fit additional data or predict future observations reliably. Thus, the statistical approach in these cases is likely to have a much poorer performance on future AOD data.

The satellite-derived PM$_{2.5}$ estimates from the CTM-based method tended to have low correlation ($R < 0.5$) with the observed daily average PM$_{2.5}$ concentrations. The highest R value obtained using a CTM-based method was in Delhi, with a value of 44 percent, but other cities had much lower R values. This low

TABLE 5.5 Correlation coefficient, mean normalized bias, and mean normalized gross error for each city

CITY	COUNTRY	LOCATION	MOST-CORRELATED SATELLITE	CORRELATION COEFFICIENT (R)	MEAN NORMALIZED BIAS	MEAN NORMALIZED GROSS ERROR
Accra	Ghana	Coastal, low altitude	Aqua	—[b]	—	—
				0.40	*−23%*	*56%*
Addis Ababa	Ethiopia	Inland, high altitude	Terra	0.63[c]	7%	24%
				0.40	*36%*	*65%*
Dakar	Senegal	Coastal, low altitude	Aqua	—[b]	—	—
				0.52	*68%*	*85%*
Delhi	India	Inland, low altitude	Terra	0.46	37%	60%
				0.44	*27%*	*60%*
Hanoi	Vietnam	Inland, low altitude	Terra	0.63	21%	43%
				0.30	*22%*	*53%*
Kampala	Uganda	Inland near lake, high altitude	Aqua	0.51[c]	6.5%	21%
				0.14	*−70%*	*70%*
Kathmandu	Nepal	Inland, high altitude	Aqua	0.66[c]	8%	24%
				0.13	*−19%*	*48%*
Lima	Peru	Coastal, low altitude	—[a]	—	—	—
Ulaanbaatar	Mongolia	Inland, high altitude	Aqua	0.44	54%	77%
				0.15	*−70%*	*78%*

Source: Original calculations for this publication.
Note: The terms correlation coefficient, mean normalized bias, and mean normalized gross error are described in the glossary of technical terms. Only results for the satellite with the highest correlation with the GLM data are presented. Statistical method results are shown in the top row of each city entry in regular type. CTM-based estimates are in the bottom row in italics. — = not available. AOD = aerosol optical depth; CTM = chemical transport model; GLM = ground-level monitoring.
a. Not available; the satellite AOD coverage was too poor to perform an evaluation.
b. Not enough GLM data were available to attempt the statistical method.
c. High *R* for the statistical method may be an artifact of the low number of points with both valid AOD and GLM data.

value of the correlation means that the CTM-based method was not able to predict more than 44 percent of the variation in daily average $PM_{2.5}$ concentrations between sites in a city or between days. This implies that current satellite methods may not be able to estimate exposures for studies of the acute health effects of $PM_{2.5}$ (which require knowledge of day-to-day variations) or of the changes in chronic exposure to $PM_{2.5}$ within a city (which requires knowledge of site-to-site variation).

Satellite AOD coverage for some cities was too sparse to allow for its use in $PM_{2.5}$ monitoring. Consistent with the results in this chapter, Lima had little valid AOD data coverage for the city. Consequently, satellite methods could not be used for this location. Ulaanbaatar had seasonal gaps in coverage, with no satellite AOD available in the winter months when $PM_{2.5}$ concentrations are the highest.

The uncertainty in all satellite-based estimates of the daily average $PM_{2.5}$ concentration at a given location in a city (as estimated with the mean normalized gross error) tended to be very large (21–77 percent for the statistical methods, and 48–85 percent for the CTM-based methods). This is consistent with the results of van Donkelaar, Martin, Brauer, and Boys (2015), who found an error in monthly average $PM_{2.5}$ concentrations over East Asia of about 50 percent. This error is much larger than usually allowed for "equivalent methods" for

FIGURE 5.3

Correlation coefficient *R* for the statistical and chemical-transport-model–based methods for the different low- and middle-income country cities

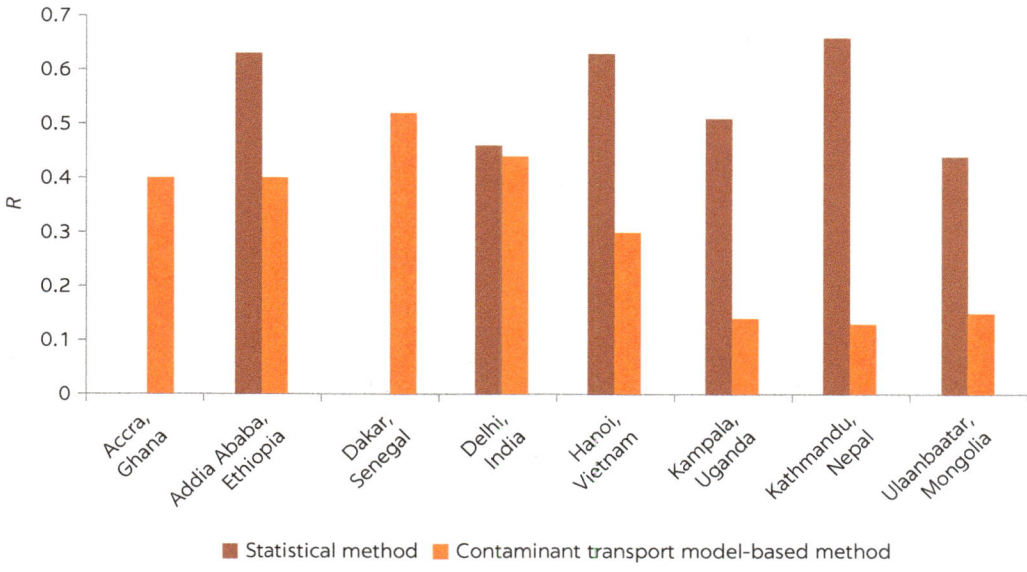

Source: Original calculations for this publication.

FIGURE 5.4

Mean normalized bias for the statistical and chemical-transport-model–based methods for the different low- and middle-income country cities

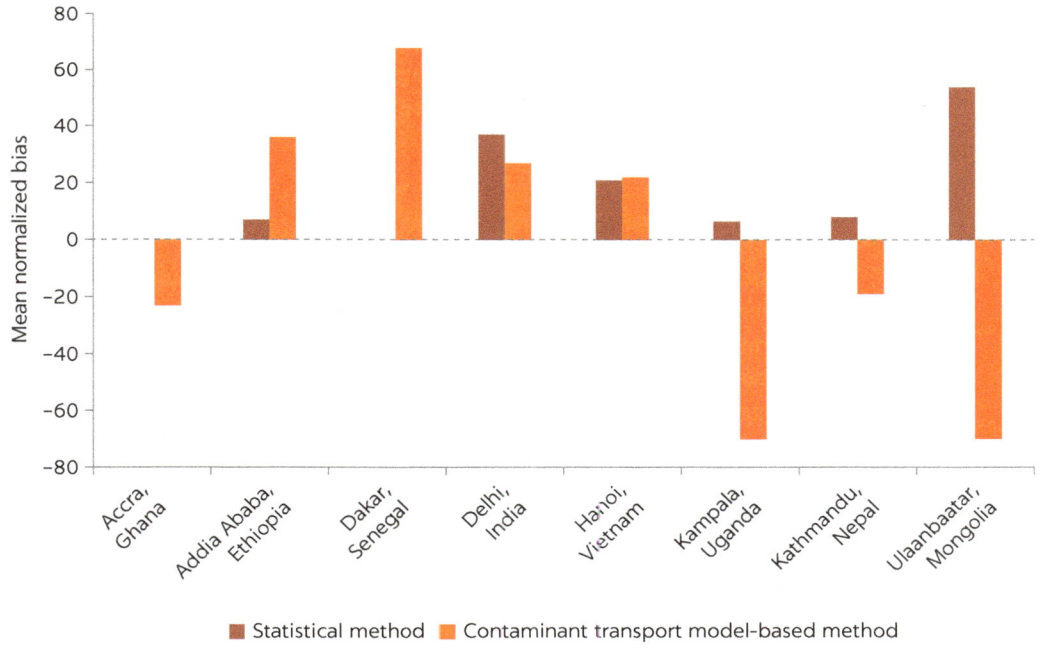

Source: Original calculations for this publication.

FIGURE 5.5

Mean normalized gross error for the statistical and chemical-transport-model–based methods for the different low- and middle-income country cities

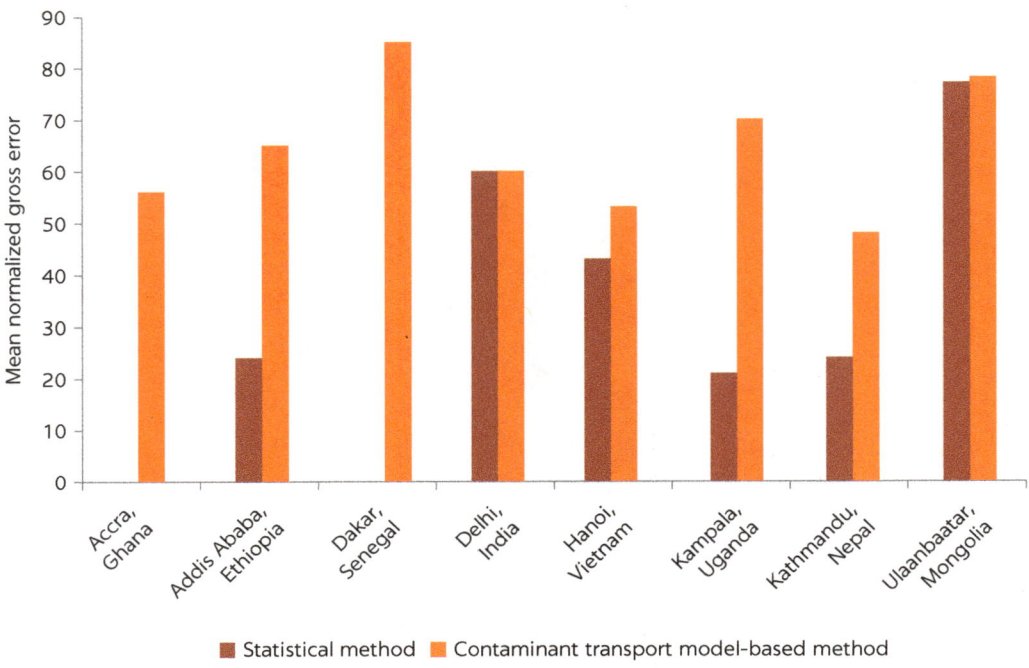

■ Statistical method ■ Contaminant transport model-based method

Source: Original calculations for this publication.

measuring PM$_{2.5}$ by the US Environmental Protection Agency (chapter 4) and thus suggests that satellite methods are not a replacement for GLM data in LMICs.

Patterns in satellite-method performance

A goal of the research presented in this report was to determine if there were patterns in the performance of satellite methods in predicting ground-level PM$_{2.5}$ concentrations based on factors such as the distance of the city from a coast, the altitude of the city, and so forth. The results in table 5.5 and appendix C do suggest some patterns may be present in the performance of satellite methods in these cities.

CTM-based methods tend to underestimate PM$_{2.5}$ concentrations in cities at high altitudes or in mountain valleys such as Ulaanbaatar (altitude 1.7 kilometers), Kathmandu (altitude 1.4 kilometers), and Kampala (altitude 1.2 kilometers). This is likely due to these cities being in small mountain valleys surrounded by rural land, and the aerosol profile from the CTM is more appropriate for rural land, which leads to an underestimate of PM$_{2.5}$ in the city (see above).

CTM-based methods may overestimate PM$_{2.5}$ concentrations in cities where desert dust is a large fraction of the PM$_{2.5}$. Although Addis Ababa is a high-altitude city (altitude 2.4 kilometers), the CTM-based method tends to overestimate in this city. This may be due to this high-altitude city being near the Sahara

Desert, and the dust from this area could be leading to the CTM overestimate in this city. Dust makes up more than half of the CTM estimate of PM$_{2.5}$ in Addis Ababa and Dakar, and both show significant overestimates when the CTM-based method is used. However, Accra also has a high percentage of PM$_{2.5}$ from dust, but for this coastal city, the CTM-based method underestimates PM$_{2.5}$.

Coastal cities (including cities near large lakes) appear to have poor satellite coverage, due to either persistent clouds or the mixture of land and water surfaces in the satellite retrieval of AOD. Accra, Dakar, and Lima all had too few matching GLM and satellite AOD points to perform the statistical method, and Kampala (near Lake Victoria) had very few points as well.

Satellite-based methods appear to work best for low-altitude, inland cities such as Delhi and Hanoi, but still have significant errors (43–60 percent) in predictions of daily average PM$_{2.5}$ concentrations at sites within the city.

The performance of the satellite approaches in estimating ground-level PM$_{2.5}$ did not appear to be affected by the type of instrument used to measure GLM data. Some cities had data only from GLM instruments at US Diplomatic Posts, which all used the beta attenuation mass monitor (BAMM) method. Accra used GLM instruments based on gravimetric methods, but the performance of satellite methods in this city was not clearly better or worse than the BAMM method. Ulaanbaatar and Delhi also included the Grimm environmental dust monitor and gravimetric filters, respectively, but this did not have a noticeable impact on the performance of satellite methods.

Recommended next steps for each city

Table 5.6 summarizes the lessons learned for each city and the recommended next steps to improve the monitoring of PM$_{2.5}$ in each to better understand and reduce the health impacts of PM$_{2.5}$.

The GLM networks in each of these cities should be expanded to better characterize the variations of PM$_{2.5}$ in the cities and surrounding suburban and rural areas, which will also allow for the derivation of better statistical models for translating satellite AOD to ground-level PM$_{2.5}$ estimates. Only Delhi and Ulaanbaatar had sufficient GLM data to derive robust statistical models, although in Ulaanbaatar this was possible for only eight months of the year due to snow cover preventing the measurement of AOD by the satellites. Lima and Ulaanbaatar have sufficiently poor AOD coverage that preclude the use of satellite observations for annual PM$_{2.5}$ monitoring, although satellites may still be useful in Ulaanbaatar for the eight months each year when AOD measurements can be taken. Similarly, the problems with poor AOD coverage near coastal regions make it unlikely that satellite methods would be successful in Accra, Dakar, or Kampala.

For inland cities without persistent seasonal snow cover, satellite observations may provide some benefits to GLM monitoring, but the current CTM-based approaches tend to underestimate significantly for the high-altitude cities. Thus, for inland cities without persistent seasonal snow (Addis Ababa, Delhi, Hanoi, and Kathmandu), a SPARTAN network site (see the section "Chemical transport model–based approaches" in appendix A) should be added to the local GLM network to directly measure the relationship between PM$_{2.5}$ and AOD using ground-based instruments, which will allow this relationship to be used instead of the CTM-based relationships in these cities.

TABLE 5.6 Lessons learned and recommended next steps to improve monitoring of PM$_{2.5}$ in each city to better understand and reduce the health impacts of PM$_{2.5}$

CITY	COUNTRY	LOCATION	LESSONS LEARNED	RECOMMENDED NEXT STEPS
Accra	Ghana	Coastal, low altitude	• Insufficient GLM and AOD data for statistical approaches to be used	• Expand current GLM network and improve QA/QC practices, including public access to instrument metadata
Addis Ababa	Ethiopia	Inland, high altitude	• Low amount of GLM and AOD data may be resulting in an over-fitted statistical approach • CTM-based approach biased high, apparently due to high dust impacts	• Expand current GLM network and improve QA/QC practices, including public access to instrument metadata • Add SPARTAN network site • After the GLM network is expanded and multiple years of data are available, test combining the GLM data with the satellite data via co-kriging to see if this approach is useful for this city
Dakar	Senegal	Coastal, low altitude	• Insufficient GLM and AOD data for statistical approaches to be used • CTM-based approach biased very high	• Expand current GLM network and improve QA/QC practices, including public access to instrument metadata • Satellite approaches unlikely to be useful even with increased GLM data due to coastal location
Delhi	India	Inland, low altitude	• Satellites can provide a citywide annual average estimate, but do not represent day-to-day or site-to-site variability • Statistical and CTM-based approaches perform similarly	• Add rural sites to current GLM network to allow use of hybrid satellite approaches and land-use regression • Add SPARTAN network site
Hanoi	Vietnam	Inland, low altitude	• Statistical and CTM-based approaches perform similarly	• Expand current GLM network and improve QA/QC practices, including public access to instrument metadata • Add SPARTAN network site • After the GLM network is expanded and multiple years of data are available, test combining the GLM data with the satellite data via co-kriging to see if this approach is useful for this city
Kampala	Uganda	Inland near lake, high altitude	• Low amount of GLM and AOD data may be resulting in an over-fitted statistical approach • CTM approach biased very low	• Expand current GLM network and improve QA/QC practices, including public access to instrument metadata • Satellite approaches unlikely to be useful even with increased GLM data due to coastal location
Kathmandu	Nepal	Inland, high altitude	• Low amounts of GLM and AOD data may be resulting in an over-fitted statistical approach	• Expand current GLM network and improve QA/QC practices, including public access to instrument metadata • Add SPARTAN network site • After the GLM network is expanded and multiple years of data are available, test combining the GLM data with the satellite data via co-kriging to see if this approach is useful for this city
Lima	Peru	Coastal, low altitude	• Satellite AOD coverage is too poor, and thus satellite approaches cannot be used for PM$_{2.5}$ monitoring in this city	• Expand current GLM network and improve QA/QC practices, including public access to instrument metadata • Investigate inconsistency between local network and US Diplomatic Post data
Ulaanbaatar	Mongolia	Inland, high altitude	• Poor satellite coverage in winter means satellite approaches cannot be used for annual PM$_{2.5}$ monitoring in this city but may be useful for the eight months a year when satellite AOD measurements exist	• Expand current GLM network and improve QA/QC practices, including public access to instrument metadata • Add SPARTAN network site • Combine satellite data during the eight-month snow-free period with GLM data via co-kriging to provide more-accurate neighborhood-scale exposure estimates

Source: World Bank.

Note: AOD = aerosol optical depth; CTM = chemical transport model; GLM = ground-level monitoring; PM$_{2.5}$ = particulate matter with an aerodynamic diameter less than or equal to 2.5 microns; QA/QC = quality assurance/quality control; SPARTAN = Surface Particulate Matter Network.

NOTES

1. Applying the GWR correction does improve the GBD estimate for Ulaanbaatar, because this correction uses GLM data in other urban areas to increase the satellite-derived PM$_{2.5}$ estimate for this city. However, the corrected value still underestimates the true annual average by a factor of two.
2. Note that here the "combined" product refers to the combination of the Dark Target and Deep Blue algorithms for a given version of the MODIS instrument (either MODIS Terra or MODIS Aqua). Thus, there are separate "combined" products from MODIS Terra and MODIS Aqua.

REFERENCES

Chatfield, R., M. Sorek-Hamer, A. Lyapustin, and Y. Liu. 2017. "Daily Kilometer-Scale MODIS Satellite Maps of PM$_{2.5}$ Describe Wintertime Episodes." Paper 263193 presented at the "Air and Waste Management Association 2017 Annual Conference and Exhibition," Pittsburgh, PA, June 5–9.

Sorek-Hamer, M., A. W. Strawa, R. B. Chatfield, R. Esswein, A. Cohen, and D. M. Broday. 2013. "Improved Retrieval of PM$_{2.5}$ from Satellite Data Products Using Non-Linear Methods." *Environmental Pollution* 182: 417–23.

van Donkelaar, A., R. V. Martin, M. Brauer, and B. L. Boys. 2015. "Use of Satellite Observations for Long-Term Exposure Assessment of Global Concentrations of Fine Particulate Matter." *Environmental Health Perspectives* 123 (2): 135–43.

van Donkelaar, A., R. V. Martin, M. Brauer, R. Kahn, R. Levy, C. Verduzco, and P. J. Villeneuve. 2010. Global Estimates of Ambient Fine Particulate Matter Concentrations from Satellite-Based Aerosol Optical Depth: Development and Application." *Environmental Health Perspectives* 118 (6): 847–55.

van Donkelaar, A., R. V. Martin, R. J. Spurr, and R. T. Burnett. 2015 "High-Resolution Satellite-Derived PM$_{2.5}$ from Optimal Estimation and Geographically Weighted Regression over North America." *Environmental Science & Technology* 49 (17): 10482–91.

6 Conclusions and Recommendations

To accomplish the objectives of this report, a literature review of the approaches used to combine satellite observations with ground-level monitoring (GLM) measurements of particulate matter with an aerodynamic diameter less than or equal to 2.5 microns ($PM_{2.5}$) was conducted. The GLM data provided by US Diplomatic Posts and other sources contained in the OpenAQ database were investigated to determine the level of quality assurance and likely uncertainties in the GLM data. Several methods for using satellite aerosol optical depth (AOD) from the publicly available sources to estimate ground-level daily average $PM_{2.5}$ concentrations within cities were then tested in nine cities in low- and middle-income countries (LMICs) to look for patterns in the performance of the satellite methods with the altitude and location of the city.

Based on the results of this work, GLM and satellite data are best thought of as complements to each other. Many GLM networks could be improved by considering satellite data, but all approaches using satellite data improve as the number of high-quality GLM sites is increased. Thus, it is important that LMICs continue to support the establishment of GLM networks to measure air pollutants that cause mortality, notably fine particulate matter ($PM_{2.5}$), in regions such as Sub-Saharan Africa and other regions with many LMICs before they can take full advantage of satellite data. The GLM data must have adequate quality assurance and quality control and follow standard operating procedures to ensure the data are of sufficient quality to be used to estimate $PM_{2.5}$ exposures for health studies and to be combined with the satellite estimates.

Satellite data may be useful for estimating air quality for countries or large areas based on average estimates. However, for the purpose of protecting human health, LMICs need to prioritize establishment or strengthening of GLM networks where they are lacking or weak. These networks can measure air quality at the level where human activity is typically carried out and where people are exposed to air pollutants, notably $PM_{2.5}$, that are harmful to health and can cause death.

The next section summarizes the conclusions of this report, and the final section provides recommendations for the use of satellite data and the improvement of GLM in LMICs.

CONCLUSIONS

Literature review

The literature review covered the approaches used (and their strengths and limitations) in each of the three key steps entailed in combining satellite observations with GLM measurements of $PM_{2.5}$: (1) the retrieval of AOD from satellite observations of reflected solar radiation, (2) the translation of the vertically integrated AOD observation into an estimate of the ground-level $PM_{2.5}$ concentration, including any bias corrections, and (3) simultaneously interpolating these satellite-derived $PM_{2.5}$ estimates and the available GLM data. The major findings of the review follow:

- Although many satellite AOD products exist, only a small number of daily products are currently freely available shortly after the observation is made at a horizontal resolution of 10 kilometers or finer.
  - Some satellite AOD products that have been used in previous studies in the literature (for example, MAIAC [Multi-Angle Implementation of Atmospheric Correction] AOD) were not operational products as of the start of this work and thus are not available for all regions in near real-time. Other satellite data (for example, the Himawari-8 and -9 data) must be purchased by the user and thus may be outside the budget for most LMICs.
  - Polar-orbiting satellite instruments (for example, MODIS [Moderate-Resolution Imaging Spectroradiometer] and VIIRS [Visible Infrared Imaging Radiometer Suite]) have the advantage of daily global coverage but tend to have relatively coarse horizontal resolution (6 to 10 kilometers). Geostationary satellites (for example, GOES and MSG-SEVIRI) make several observations each day at a higher horizontal resolution (2 to 4 kilometers) but observe only one side of the globe, and their AOD retrievals tend to exclude coastal areas.
  - Accessing the VIIRS and GOES products is difficult, since the download process cannot be easily automated.
  - The MODIS combined Deep Blue and Dark Target AOD retrieval is free, is available in near real-time, is easy to access, has been extensively validated, and has good coverage over urban and coastal areas.
- Several studies have successfully converted satellite AOD to monthly average and annual average ground-level $PM_{2.5}$ estimates using statistical techniques, chemical transport model (CTM)–based approaches, or hybrid approaches with typical errors of 30–50 percent, but very few have tried to predict daily $PM_{2.5}$ concentrations or their variation within a city.
  - CTM-based approaches are easy to implement globally and do not require any GLM data. Consequently, CTM-based approaches can be implemented for all LMICs. However, these approaches can have significant biases due to errors in the CTM's simulation of the vertical profile of the aerosols, as well as in the mismatch between the horizontal resolution of the CTM and the distribution of population within a city.
  - Statistical approaches are locally unbiased (since these methods train a statistical model to minimize the mean bias) and do not depend on an initial estimate of the aerosol vertical profile. However, statistical approaches require a multiyear GLM data record with concurrent satellite observations to train the statistical model. Only a few LMICs

(for example, India or Mongolia) have sufficient GLM data to use these approaches.

- Hybrid approaches that correct CTM-based approaches with statistical models trained with GLM data tend to give the best performance. However, they also require a multiyear GLM data record with concurrent satellite observations over many different land-use types (for example, urban and rural sites or coastal and inland sites) before the statistical model can be trained.
- Co-kriging and land-use regression (LUR) can be used to combine satellite-derived $PM_{2.5}$ estimates with GLM data to provide estimates of $PM_{2.5}$ concentrations at a neighborhood scale across urban areas.
  - Co-kriging can be used to estimate variations in daily average $PM_{2.5}$ concentrations, whereas LUR can be used for annual or longer-term averages.
  - Co-kriging and LUR tend to work better as the satellite horizontal resolution is increased and as the GLM network covers a wider variety of sites (for example, urban and rural).

Quality assurance of ground-level monitoring data in LMICs

All the cities evaluated in this project would benefit from expanding their GLM networks and enhancing their quality-assurance procedures, and this should be a key next step in monitoring $PM_{2.5}$ in these cities.

Missing or unreliable GLM data prevent the use of statistical or hybrid approaches to incorporating satellites observations into $PM_{2.5}$ monitoring. When GLM data are absent, only the CTM-based approaches can be applied. When GLM data are present but of poor quality (that is, unreliable), a bias correction can be calculated for the CTM-based approaches, but there is no assurance that the "corrected" satellite value is more accurate than the original estimate. Finally, poor maintenance of GLM instruments can lead to large data gaps, which either prevent the use of statistical methods or bias the statistical models trained with the data.

Most $PM_{2.5}$ instruments in use around the world have been approved by the US Environmental Protection Agency (EPA) as federal equivalent methods (FEMs) for $PM_{2.5}$ characterization and thus can provide reliable measurements of ground-level $PM_{2.5}$ if they are properly maintained and calibrated. To qualify as a FEM, the $PM_{2.5}$ measurement technique must demonstrate uncertainties less than 10 percent relative to the federal reference method for 24-hour average $PM_{2.5}$ concentrations. However, the actual accuracy and precision of the data will depend on whether proper maintenance and procedures are followed, and if the instruments are monitored for errors.

US Diplomatic Posts are equipped to measure $PM_{2.5}$ with Met One beta attenuation monitor (BAM) systems and abide by the current EPA guidelines for $PM_{2.5}$ monitoring. Thus, these posts are sources of high-quality GLM data for the locations where they operate. However, these data may not be representative of average air quality in the city where these posts are located. Whereas the Delhi, India, local government GLM network appears to be consistent with the US Diplomatic Post data, the local network in Lima, Peru, is inconsistent. More details on the maintenance and quality-assurance procedures in the GLM network in Lima would be needed to determine the causes of the observed differences.

Application of satellite approaches to cities in LMICs

The results of the application of satellite approaches to cities in LMICs suggest that satellites cannot be a replacement for a high-quality GLM network. This is because several limitations to the use of satellite AOD are found in the cities tested:

- Coastal cities (Accra, Dakar, Kampala, and Lima) and cities with persistent snow cover (Ulaanbaatar) have significant limitations in satellite AOD coverage.
  - Satellite coverage was poorest in the coastal city of Lima, and thus an evaluation of CTM-based approaches with GLM network data was not possible. The other coastal cities (including cities near large lakes) examined in this study had poor satellite coverage, because of either persistent clouds or the mixture of land and water surfaces in the satellite retrieval of AOD.
  - In Ulaanbaatar, no satellite AOD products are available for the high $PM_{2.5}$ winter months of December to mid-March.
- The ground-level $PM_{2.5}$ estimates from the CTM-based and statistical approaches for converting satellite AOD generally had a low correlation with the true daily average $PM_{2.5}$ values within a city over the year.
- The uncertainty (that is, the average error that is expected for a single estimate) in all satellite-based estimates of the daily average $PM_{2.5}$ concentration at a given location in a city tended to be very large (21–77 percent for the statistical methods, and 48–85 percent for the CTM-based methods), and thus these methods cannot be considered "equivalent methods" when compared with United States– or European Union–approved GLM techniques.[1]
- CTM-based satellite approaches tend to underestimate $PM_{2.5}$ in high-altitude cities. For example, annual average $PM_{2.5}$ concentrations in Ulaanbaatar were underestimated by a factor of 10.
- Estimating variations in annual average $PM_{2.5}$ within a city (to estimate chronic health effects on a neighborhood scale) is unlikely to be possible with satellite AOD data using the approaches tested here. This will require GLM-network data and LUR, with the satellite AOD product or the satellite-based $PM_{2.5}$ estimate used as a variable in the LUR.
- Both the CTM-based and statistical approaches tested here showed little ability to represent the day-to-day variability in $PM_{2.5}$ concentrations, with average absolute errors of ±50 percent for the best approaches within each city. Thus, studies of acute health effects will likely require GLM data.

However, a few cases and conditions are found where the use of satellite data may be appropriate as a complement (but not a replacement) for GLM networks. For example, satellite-based methods appear to work best for low-altitude, inland cities such as Delhi and Hanoi but still have significant errors (43–60 percent) in predictions of daily average $PM_{2.5}$ concentrations at sites within these cities. In addition, the results suggest that under some conditions, adding satellite data to GLM network data via co-kriging may reduce the number of GLM sites needed to characterize $PM_{2.5}$ concentrations and exposure within a city, although this ability varies from city to city and could lead to large errors in annual estimates for cities with a seasonal or persistent lack of satellite AOD coverage.

RECOMMENDATIONS

Use of satellite data in PM$_{2.5}$ monitoring

Table 6.1 summarizes typical problems associated with the use of satellite data for PM$_{2.5}$ monitoring in LMICs, the consequence of the problem, and the recommended action to address each problem.

Inland cities without persistent seasonal snow cover may be able to benefit from the use of satellite observations in their PM$_{2.5}$ monitoring, and so a key next step should be establishing a SPARTAN network site within these cities to directly measure the relationship between AOD and ground-level PM$_{2.5}$ concentrations, thus reducing the biases that come from using CTM-based estimates of this relationship.

The research conducted for this report also suggests the following recommendations for the use of satellite observations to supplement GLM data in LMICs based on the typology proposed in chapter 1 (see also table 6.2):

- For Type I countries, since no GLM data exist, the only possible approach to convert AOD to ground-level PM$_{2.5}$ concentrations for these countries is a CTM-based approach. Thus, the raw CTM-based estimates of ground-level PM$_{2.5}$ concentrations at the native resolution of the satellite AOD product should be used. However, the derived ground-level PM$_{2.5}$ values should be

TABLE 6.1 **Summary of typical problems in using satellites in PM$_{2.5}$ monitoring, the consequences, and the recommended actions to further the use of satellite data in PM$_{2.5}$ monitoring to better understand and reduce the health impacts of PM$_{2.5}$ in low- and middle-income countries**

PROBLEM	CONSEQUENCE	RECOMMENDATION
Lack of GLM data (Type I LMICs).	Only CTM-based approaches can be used to monitor PM$_{2.5}$ with satellites, and accuracy cannot be verified.	Install additional GLM sites and follow rigorous quality-assurance and maintenance procedures.
GLM data exist but are of poor quality (Type II LMICs).	Only CTM-based approaches can be used to monitor PM$_{2.5}$ with satellites, and accuracy cannot be verified.	Establish and follow rigorous quality-assurance and maintenance procedures for GLM data and report GLM metadata with actual measurements.
City is surrounded by rural land within typical resolution of global chemical transport models (that is, cities in mountain valleys).	CTM-based aerosol profile is incorrect, biasing satellite results.	Establish a SPARTAN network site to directly measure AOD and PM$_{2.5}$ relationship.
City has seasonal persistent snow or cloud cover.	Satellite estimates will be unavailable during key seasons, biasing annual averages.	Satellite approaches should not be used for annual averages but may be used in certain seasons.
City is near a major body of water.	Many AOD products will have poor coverage, since they filter out coastal regions.	Satellite coverage for city should be assessed on a case-by-case basis.
City has only urban GLM sites.	Land-use-regression and hybrid approaches are not useful because of the lack of data on suburban and rural PM$_{2.5}$ concentrations.	Establish a few nearby rural GLM sites to provide data on urban-rural variation of PM$_{2.5}$.

Source: World Bank.
Note: AOD = aerosol optical depth; CTM = chemical transport model; GLM = ground-level monitoring; LMIC = low- and middle-income country; PM$_{2.5}$ = particulate matter with an aerodynamic diameter less than or equal to 2.5 microns, SPARTAN = Surface Particulate Matter Network.

TABLE 6.2 Recommendations for the use of satellite observations to supplement ground-level monitoring data in low- and middle-income countries

COUNTRY TYPE	RECOMMENDATIONS
Type I: No air-quality data	• The raw CTM-based estimates of ground-level $PM_{2.5}$ concentrations at the native resolution of the satellite AOD product should be used. • The satellite estimates should be assigned a high uncertainty (at least 50 percent).
Type II: Countries with some air-quality data but of variable quality without rigorous QA procedures	• Both the raw and bias-corrected CTM-based estimates of ground-level $PM_{2.5}$ concentrations at the native resolution of the satellite AOD product should be reported and stored. • The satellite estimates should be assigned a high uncertainty (at least 50 percent). • As quality-assurance procedures improve (Type III), the satellite bias corrections should be recalculated.
Type III: Countries that possess reliable air-quality information but with poor spatial or temporal coverage	• Both statistical and bias-corrected CTM-based approaches for converting AOD to $PM_{2.5}$ should be tested, and the most reliable approach for a given city selected. • Test combining these satellite estimates with GLM data via LUR and co-kriging.
Type IV: Good, reliable air-quality monitoring underway or being established	• Both statistical and bias-corrected CTM-based approaches for converting AOD to $PM_{2.5}$ should be tested, and the most reliable approach for a given city selected. • Test combining these satellite estimates with GLM data via LUR and co-kriging.
Type V: Routine, long-term air-quality monitoring	• Both statistical and bias-corrected CTM-based approaches for converting AOD to $PM_{2.5}$ should be tested, and the most reliable approach for a given city selected. • Test combining these satellite estimates with GLM data via LUR and co-kriging.

Source: World Bank.
Note: AOD = aerosol optical depth; CTM = chemical transport model; GLM = ground-level monitoring; LUR = land-use regression; $PM_{2.5}$ = particulate matter with an aerodynamic diameter less than or equal to 2.5 microns; QA = quality assurance.

assigned a high uncertainty that reflects not only the uncertainty in the AOD but also the estimated uncertainty in the CTM-derived AOD to $PM_{2.5}$ relationship (for example, about 50 percent based on van Donkelaar and others 2015 or the 22–85 percent errors found by the research conducted for this report).
• For Type II countries, the small amount of GLM data (with variable quality) that exists will provide at least some ability to derive a bias estimate for the raw CTM-based estimates of ground-level $PM_{2.5}$ concentrations discussed above. However, without improvement in the quality-assurance procedures in these countries, it will not be possible to be certain that the bias-corrected estimate is truly more accurate than the raw estimate. Thus, it is advisable to report and store both values. This will allow for reprocessing of the satellite estimates when more-rigorous quality-assurance procedures are developed.

- The GLM data in Type III countries potentially allow both statistical and bias-corrected CTM-based approaches for converting AOD to $PM_{2.5}$ to be considered. CTM-based approaches likely would be the best approach for these countries, but statistical approaches should also be tested before making a final decision. LUR can be combined with either technique to provide a finer-scale estimate of annual average ground-level $PM_{2.5}$ concentrations (at about 0.5-kilometer resolution), and co-kriging can be used to derive high-resolution daily estimates across the city.

- Type IV and V countries will be able to take advantage of the same approaches outlined above for Type III countries to fill the gaps in the existing GLM network to help cover more of the country's population. The more extensive GLM networks in these countries will allow for more-accurate estimates of the geographical and seasonal variation in the AOD to $PM_{2.5}$ relationship. Thus, at this point, purely statistical approaches may begin to outperform CTM-based estimates, especially for regions where the model's emission inventories are out-of-date or otherwise inaccurate.

Presented below are this report's recommendations on data sources and other methodological concerns for LMICs interested in using satellite data in their $PM_{2.5}$ monitoring and exposure assessment:

- Since the uncertainties of the different standard satellite AOD products are similar, all countries should use the satellite product that offers the best balance of satellite AOD coverage and fine horizontal resolution for their country. The highest horizontal resolution possible for satellite AOD will likely be between 2 and 4 kilometers (for geostationary observations) or 1 and 10 kilometers (for polar observations). Consequently, finer-scale predictions will be possible only for areas where extensive, high-quality GLM data exist.

- The satellite AOD coverage, both spatially and seasonally, should be assessed for the country before any attempt to integrate satellite data is performed. The physical limits of satellite AOD coverage will make their use in $PM_{2.5}$ monitoring difficult in many cities.

  - In cities with persistent wintertime snow cover, such as Ulaanbaatar, or seasonal persistent cloud cover, such as in Lima, all approaches using satellite AOD will likely not give accurate annual averages from current satellites.

  - In coastal cities, the mix of water and land surfaces within a satellite footprint and persistent clouds may also mean that using satellite AOD is not an option.

- For countries that do not have the capacity to perform their own meteorological or chemical transport modeling, the freely available global model data sets provided by organizations in the United States for estimates of planetary boundary layer height, aerosol vertical profiles, and other parameters can be used in a CTM-based approach to estimating ground-level $PM_{2.5}$ from satellite measurements. For long-term studies, the National Aeronautics and Space Administration's MERRA-2 reanalysis can be used, because this will represent the best estimate of the historical atmospheric state. However, the reanalysis takes two months to produce, and thus for short-term forecasts and advisories, the National Centers for Environmental Prediction's Global Forecast System (NCEP GFS) output (including the NEMS GFS Aerosol Component aerosol forecasts) can be used. The initial tests suggest that these data sets perform like the custom model data sets used in the GBD 2016 database for annual averages.

- Purchasing modeling data sets with higher resolution than the MERRA-2 and NCEP GFS products could address the errors seen in CTM-based methods for cities in appreciably different air-quality environments than their surroundings, such as cities in mountain valleys surrounded by rural land such as Ulaanbaatar.

Quality-assurance procedures for ground-level data

Based on the research conducted in preparing this report, it is important that LMICs continue to support the establishment of GLM networks in regions such as Sub-Saharan Africa and other regions where GLM networks are weak or non-existent. These GLM data must have adequate quality assurance and quality control and follow standard operating procedures to ensure the data are of sufficient quality.

In addition, the results of the research conducted for this report suggest the following recommendations for LMICs and the US State Department for the quality-assurance procedures and the reporting of metadata for their GLM measurements:

- $PM_{2.5}$ measurements need to be provided along with information on the instrument/technique type, estimates of measurement uncertainties, relevant metadata, and operational history. It is important that each $PM_{2.5}$ measurement not only identifies the type of instrument used for the measurement, but also tracks the instrument's relevant history of use and calibration. This includes all "meta" data from instrument data files, including "housekeeping" files.
- Any recorded results from standard operating procedure and quality-assurance procedures tests should be collected and disseminated along with the $PM_{2.5}$ measurements to establish an instrument track record over time providing a transparent, data-centric measure of instrument reliability and performance.
- Currently, US embassies do not publish their full quality-assurance protocol metadata along with the $PM_{2.5}$ levels, thereby reducing the use of these measurements as well-established and controlled reference measurements. Thus, it is recommended that US embassies publish their full quality-assurance protocol and relevant metadata along with the $PM_{2.5}$ levels.
- Near real-time GLM data should be assessed using automated scripts such as the OpenAQ quality wrapper developed in this project to identify potential problems quickly so they can be addressed. This will help focus the time of air-quality staff in LMICs, allowing for higher-quality data while keeping the required number of trained technicians small.

NOTE

1. These percentages are the mean normalized gross error (average of the absolute value of the differences between the satellite estimate and the observations divided by the observations) for satellite-based predictions of daily average $PM_{2.5}$ concentrations at each GLM site within a city and thus are an estimate of the error that could be expected when using the satellite methods to predict daily average $PM_{2.5}$ concentrations at a location within the city that does not include a GLM site.

REFERENCES

GBD 2016 Risk Factors Collaborators. 2017. "Global, Regional, and National Comparative Risk Assessment of 84 Behavioural, Environmental and Occupational, and Metabolic Risks or Clusters of Risks, 1990–2016: A Systematic Analysis for the Global Burden of Disease Study 2016." *Lancet* 390: 1345–422.

van Donkelaar, A., R. V. Martin, M. Brauer, and B. L. Boys. 2015. "Use of Satellite Observations for Long-Term Exposure Assessment of Global Concentrations of Fine Particulate Matter." *Environmental Health Perspectives* 123 (2): 135–43.

APPENDIX A

Literature Review

Current approaches to converting total column aerosol optical depth (AOD) retrieved from satellites into estimates of ground-level $PM_{2.5}$ (particulate matter with an aerodynamic diameter less than or equal to 2.5 microns) concentrations have various limitations. First, retrieved total column AODs from satellites have errors associated with some issues. Such issues include excessive spatial averaging that mixes urban and rural areas and thus underestimates AOD in the more polluted urban areas (for example, Hu, Waller, Lyapustin, Wang, Al-Hamdan, and others 2014; Hu, Waller, Lyapustin, Wang, and Liu 2014; Lyapustin and others 2011); reflective surfaces, such as desert dust, that make it more difficult to separate the reflection of sunlight by aerosols from the surface reflection (for example, Sorek-Hamer and others 2015); "patchwork" surfaces in urban areas, where several different land-use types, each with different albedos, may all be present within a single satellite pixel (for example, Oo and others 2010); and the presence of undetected thin clouds, which can also create a positive bias in AOD retrievals (for example, Sun and others 2011). These limitations in satellite AOD products are consistent with some of the errors identified in using satellite observations to determine ground-level $PM_{2.5}$ concentrations. For example, the high estimates in Egypt, Qatar, and Saudi Arabia are consistent with overestimates of AOD in regions with highly reflective surfaces. These problems can be mitigated by careful analysis and identification of the satellite observations and retrieval algorithms that give the most accurate and precise estimates of AOD for a given region, rather than trying to find one satellite product that works sufficiently well for all conditions around the globe. A deep understanding of the strengths and weaknesses of the different satellite AOD products is thus essential to determining which product will be most useful for a given set of geographic and meteorological conditions.

Most previous validation work on the use of satellite observations to estimate ground-level $PM_{2.5}$ has been performed in developed countries with extensive, well-calibrated, long-term ground-level-monitoring (GLM) observations of $PM_{2.5}$ (Type V countries, using the proposed typology in chapter 1). However, the performance of the statistical models and chemical transport models (CTMs) used to derive these relationships is not as well characterized in many low- and middle-income countries (LMICs), where GLM is infrequent or absent and few field campaigns have been performed to make $PM_{2.5}$ observations with which to train and test the models. In addition, any GLM observations that are available in LMICs may be difficult to access or may require significant effort to quality-assure the data. Being able to access these data sets and assess the quality of the measurements is thus a critical part of this project. The lack of GLM in many LMICs also introduces difficulties in trying to relate the spatially averaged AOD observations to variations in $PM_{2.5}$ within a city (due to distance from roadways and other pollution sources), which suggests the need to combine satellite observations with GLM observations and data on pollution sources within a city using interpolation techniques (for example, de Hoogh and others 2016; Vienneau and others 2013).

This appendix contains a literature review that discusses the approaches used (and their strengths and limitations) in each of the three key steps in combining satellite observations with GLM measurements of $PM_{2.5}$:

- The retrieval of AOD from satellite observations of reflected solar radiation (second section)
- The translation of the vertically integrated AOD observation into an estimate of the ground-level $PM_{2.5}$ concentration, including any bias corrections (third section)
- Simultaneously interpolating these satellite-derived $PM_{2.5}$ estimates and the available GLM data (fourth section).

Based on this review, recommendations are made for LMICs on how best to incorporate satellite data into their monitoring plans (fifth section). The recommendations use the following proposed typology of countries based on their levels of engagement with air-quality monitoring:

- Type I: Countries with no existing measurements and no history of routine measurements of atmospheric composition of any kind. Some anecdotal measurements or one-time sampling may have taken place.
- Type II: Countries with some information on atmospheric composition available (perhaps PM_{10} or total suspended particulates [TSPs]) but of variable quality without rigorous quality assurance (QA) procedures.
- Type III: Countries that possess reliable information but with poor spatial or temporal coverage. For example, monitoring may exist in only one city or routine monitoring may exist for a period of a year or two but is no longer being collected because of equipment malfunction or lack of repairs.
- Type IV: Good, reliable AQ monitoring underway or being established.
- Type V: Routine, long-term AQ monitoring.

AVAILABLE AEROSOL-OPTICAL-DEPTH PRODUCTS

Monitoring the distribution and properties of atmospheric aerosols from satellites is a crucial component of establishing the radiative forcing of aerosols and their impact on climate, and it is of growing interest for regional and urban air-quality monitoring. The degree to which aerosol particles suspended in the atmosphere interact with light depends on the size and shape of the particles. Smaller particles that scatter light at visible (VIS) and shorter ultraviolet (UV) wavelengths have negligible effect at longer near-infrared (NIR) wavelengths where (in the absence of clouds) there is a relatively clear view to the surface.

These dependencies on spectral properties can be exploited through satellite observations in the VIS through NIR wavelengths to infer properties of the aerosols classified according to a limited set of models. The ability to measure aerosols is dependent on there being reasonably good contrast between the light scattered from the particles and that reflected from the surface at VIS wavelengths. For this reason, aerosol retrievals are often limited to darker surfaces (for example, ocean or vegetation). Integral to this approach is the need to infer the surface reflectance at VIS wavelengths in the absence of aerosols. This is typically achieved either by using observations from previous times (when aerosols were not present) or by relying on relationships with the signals at longer wavelengths unaffected by the aerosols. These methods tend to also work best for darker surfaces. The monitoring of aerosols based on reflected light is also restricted to daytime conditions.

Radiative-transfer (RT) models enable accurate simulations of reflectance at the top of the atmosphere (TOA) in the presence of aerosol layers with a variety of optical properties. By properly accounting for surface reflectance from land and ocean backgrounds, it is possible to retrieve several important aerosol properties by comparing satellite-observed TOA reflectances to those calculated from an RT model. Typically, to optimize processing time, TOA reflectances are precalculated by the RT model for several combinations of aerosols and are stored in look-up tables (LUTs) accessed by a retrieval algorithm. The RT calculations are performed for a range of aerosol optical thicknesses, so each stored reflectance value in the LUT corresponds to an aerosol optical thickness (AOT).

Because the surface type plays such an important role in aerosol retrievals, and because concentrations of aerosols often correspond to land or ocean provenance regions, somewhat different approaches are implemented over land and ocean. For land retrievals, specification of land surface reflectance is more difficult because of the greater heterogeneity of global land cover. For this reason, direct observations of land-surface reflectance are preferred over using simulated land-surface reflectances. Threshold tests are typically applied to all land pixels to ensure that only dark vegetated pixels are processed. The total TOA reflectance is the sum of the observed land surface and atmospheric contributions. The matching aerosol model is one with a TOA reflectance that produces the smallest residual with respect to observed TOA reflectances in bands sensitive to the presence of aerosols. The AOD and other properties associated with the selected model are reported.

Variations on this retrieval approach have been used to generate aerosol products from satellite observations. The World Meteorological Organization (WMO) maintains an online record of satellite technology and products related to weather, water, and climate called the Observing Systems Capability Analysis and Review Tool (OSCAR)[1] that is a very good reference of all past, current, and future aerosol products. The operational and science remote sensing capabilities collected there provide different levels of information about the aerosol constituents. Multispectral imaging radiometers (for example, Advanced Baseline Imager [ABI], Moderate-Resolution Imaging Spectroradiometer [MODIS], and Visible Infrared Imaging Radiometer Suite [VIIRS]) are capable of column AOD and aerosol particle size determinations. Multiangle radiometers (for example, Multi-angle Imaging Spectroradiometer [MISR]) and polarization imagers (for example, Multi-viewing, Multi-channel, Multi-Polarization Imaging mission [3MI] or Polarization and Directionality of the Earth's Reflectances [POLDER]) can provide more detail on aerosol size or type information, and space-based lidars (light detection and ranging devices) (for example, Cloud-Aerosol Lidar with Orthogonal Polarization [CALIOP]) are used for aerosol and cloud profiling.

Aerosol products derived from instruments aboard polar satellites in low earth orbit (LEO) (for example, Advanced Very High Resolution Radiometer [AVHRR], MODIS, and VIIRS) provide global coverage roughly once per day, with spatial resolution on the order of 1 to 10 kilometers, and that depends on the orbit path. Products derived from instruments on geostationary (GEO) satellites (for example, ABI, Advanced Himawari Imager [AHI], Geostationary Operational Environmental Satellite [GOES-Imager], and Spinning Enhanced Visible and Infrared Imager [SEVIRI]) provide coverage over the full field of view (that is, Full Disk) at less than 30-minute frequency with a spatial resolution of roughly 2 to 4 kilometers, and that varies with the satellite zenith angle. As a resource for air-quality monitoring at specific locations around the world, it is necessary that remotely sensed aerosol-optical-depth products used to infer ground-level PM2.5 concentrations are available in near real-time (NRT) and that the products provide adequate coverage for regions of interest. Figure A.1 illustrates the coverage provided by LEO and GEO satellites in the northern hemisphere. A summary of the aerosol products produced by the current series of LEO and GEO satellites is provided in table A.1.

Satellite-based AOD products typically undergo a very rigorous validation process before they are officially released. This process includes extensive comparisons to ground-based AOD measurements (for example, Aerosol Robotic Network [AERONET]; Holben and others 2001) as well as intercomparisons with other established AOD products and with other science missions that can provide more detailed aerosol information (for example, aerosol profiles via CALIOP). As a result, the errors are fairly well documented and are typically expressed as a bias from truth (that is, accuracy) and the scatter relative to this bias (that is, the precision), though the metrics used can vary from program to program. In addition, products are always accompanied by data quality flags that identify locations of good and degraded performance with strong recommendations to refer to these ratings when used in any application. For $PM_{2.5}$ air-quality applications, these error specifications should be used to propagate uncertainties in the satellite retrievals of AOD to the estimates of ground-level $PM_{2.5}$ concentrations.

The next two subsections provide an overview of satellite-based AOD retrievals from polar and geostationary platforms. One significant finding of this review is that alternative products have sometimes been developed by different organizations, but not all AOD products described in the published literature are currently made available to users. The National Oceanic and Atmospheric Administration (NOAA) MODIS products are currently the most accessible, but the long-term status of MODIS is uncertain, with the Joint Polar Satellite System (JPSS) VIIRS instrument expected to provide continuing coverage into the future. The AOD products from GEO satellites can provide greater temporal coverage of aerosol events, but currently the only reliable coverage is from the the European Organisation for the Exploitation of Meteorological Satellites (EUMETSAT) Meteosat Second Generation (MSG) satellite providing coverage of Europe and Africa. The Geostationary Operational Environmental Satellite-R

FIGURE A.1

Coverage provided by available polar and geostationary weather satellites

a. Current coverage provided by available polar and geostationary satellites

b. Future coverage after planned updates of geostationary satellites

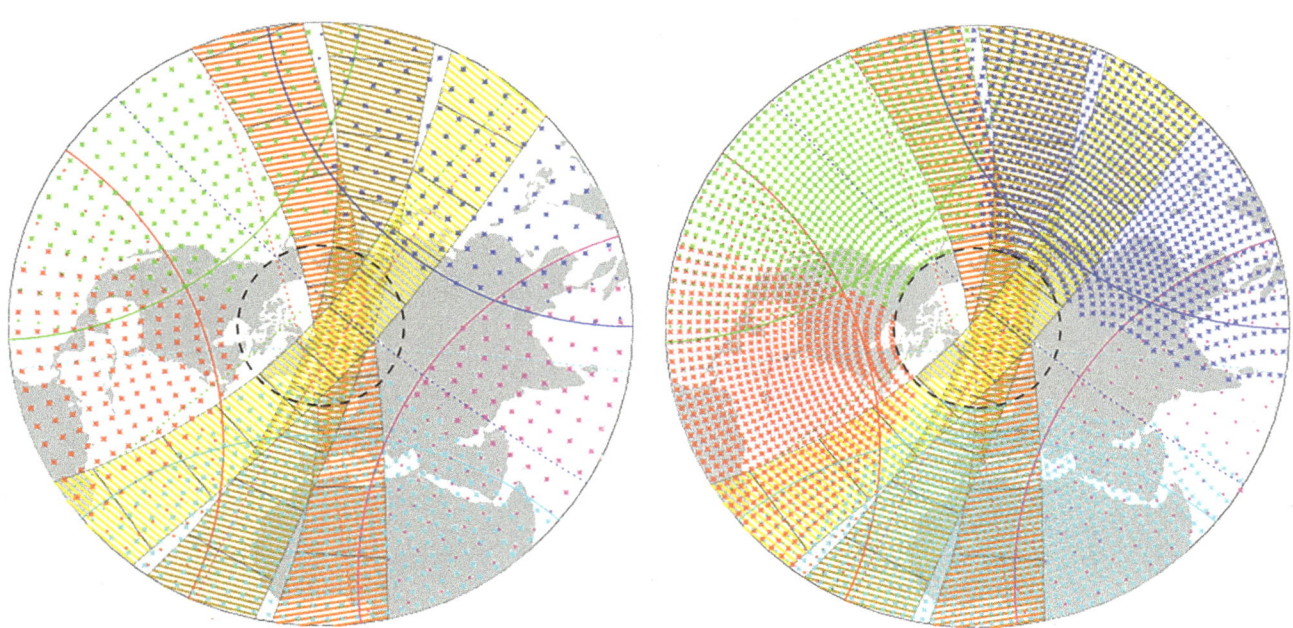

Source: World Bank.
Note: The three bands in each panel illustrate three consecutive swaths from the Moderate-Resolution Imaging Spectroradiometer in sun-synchronous orbit providing once per day coverage. Green, red, turquoise, magenta, and blue points represent coverage from the geostationary satellites. The left panel shows current coverage; the right panel shows future coverage with roughly two times higher spatial resolution.

TABLE A.1 **Summary of operational aerosol products**

AGENCY [DEVELOPER]	SATELLITE [INSTRUMENT]	ORBIT [TYPE]	PRODUCT(S)	REFRESH [COVERAGE]	NEAR REAL-TIME AVAILABILITY
EUMETSAT [CNRS-ICARE]	MSG [SEVIRI]	0 E [GEO]	SMAOL	15 minutes [Europe, Africa]	Yes
EUMETSAT [EUMETSAT]	Metop-A Metop-B [AVHRR, GOME-2, IASI]	9:30 dsc 9:30 dsc [LEO]	Polar Multi-Sensor Aerosol Product HSR 5 × 40 10 × 40 kilometer swath 960 × 1920 kilometer	Daily [Global]	Yes
JMA [JMA-MSC]	Himawari-8/9 [AHI]	141. E [GEO]	AOD	[East Asia, Indonesia, Australia]	TBD
NASA [NASA-GSFC]	Terra Aqua [MODIS]	10:30 dsc 13:30 asc [LEO]	10 km Dark Target AOD 10 km Deep Blue AOD 10 km combined AOD 3 km Dark Target AOD	Daily [Global]	Yes
NASA [NASA-GSFC]	Terra Aqua [MODIS]	10:30 dsc 13:30 asc [LEO]	MAIAC AOD (1 kilometer) gridded Amazon data set	Daily [Global]	TBD 2000–12

continued

TABLE A.1, *continued*

AGENCY [DEVELOPER]	SATELLITE [INSTRUMENT]	ORBIT [TYPE]	PRODUCT(S)	REFRESH [COVERAGE]	NEAR REAL-TIME AVAILABILITY
NOAA [NASA-GSFC]	NOAA-20 [VIIRS]	13:30 asc [LEO]	10 kilometer Dark Target AOD 10 kilometer Deep Blue AOD 10 kilometer Combined AOD 3 kilometer Dark Target AOD	Daily [Global]	TBD
NASA [JPL]	Terra [MISR]	10:30 dsc [LEO]	AOD (MIL2ASAE)	9 days [Global]	Yes
NOAA [NOAA-NCEI]	NOAA-19 NOAA-15 NOAA-18 [AVHRR]	15:20 asc 18:15 asc 19:02 asc [LEO]	AOT	Daily [Global; Ocean only]	Yes
NOAA [NOAA-SPSD]	GOES-13 GOES-15 [IMAGER]	75. W 135. W [GEO]	GASP-East AOD GASP-West AOD	1 hour [CONUS]	Yes
NOAA [NOAA-STAR]	NOAA-20 [VIIRS]	13:30 asc [LEO]	AOT	Daily [Global]	Yes
NOAA [NOAA-STAR]	GOES-16 GOES-17 [ABI]	75. W 137. W [GEO]	SMAOD	5, 10, or 15 minutes [North and South America]	Yes

Source: World Bank.
Note: In column 3, Low Earth orbits (LEOs) are all sun-synchronous and identified in terms of the daytime equator crossing time and direction (that is, ascending [asc] or descending [dsc]); Geostationary (GEO) orbits are identified by their satellite longitude. ABI = Advanced Baseline Imager; AOD = aerosol optical depth; AVHRR = Advanced Very High Resolution Radiometer; CNRS-ICARE = Centre National de la Recherché Scientific–Institut de Combustion, Réactivité et Environnement; CONUS = continental United States; EUMETSAT = European Organisation for the Exploitation of Meteorological Satellites; GASP = GOES Aerosol/Smoke Product; GOES = Geostationary Operational Environmental Satellite; GOME-2 = Global Ozone Monitoring Experiment-2; GSFC = Goddard Space Flight Center; IASI = Infrared Atmospheric Sounding Interferometer; IMAGER = GOES Imager; JMA = Japan Meteorology Agency; JPL = Jet Propulsion Laboratory; MAIAC = Multi-Angle Implementation of Atmospheric Correction; MIL2ASAE = MISR level 2 aerosol parameters; MISR = Multi-angle Imaging Spectroradiometer; MODIS = Moderate-Resolution Imaging Spectroradiometer; MSC = Meteorological Satellite Center; MSG = Meteosat Second Generation; NASA = National Aeronautics and Space Administration; NOAA = National Oceanic and Atmospheric Administration; SEVIRI = Spinning Enhanced Visible and Infrared Imager; SMAOD = suspended matter, aerosol optical depth; SMAOL = SEVIRI-MSG Aerosol Over Land; STAR = Center for Satellite Applications and Research; TBD = to be determined; VIIRS = Visible Infrared Imaging Radiometer Suite.

series (GOES-R) products over North and South America were made available in 2018. However, the availability of Japan Meteorology Agency (JMA) Himawari data with coverage over East Asia and Indonesia is uncertain, and no current geostationary satellite mission is providing AOD products in the region of India and western Asia. Furthermore, the availability of satellite coverage does not ensure that AOD products will be produced, because the retrievals are limited by environmental conditions (for example, must be cloud-free) and surface conditions (for example, must be snow-free). The third subsection looks at three urban locations to review how satellite-derived AOD coverage might be provided to support air-quality missions.

Polar-orbiting satellites

MODIS products

Aerosol products derived from observations by the MODIS aboard the National Aeronautics and Space Administration's (NASA's) Terra and Aqua satellites are among the most widely studied products in the aerosol remote sensing community. This pair of instruments (MODIS-Terra and MODIS-Aqua) has created a climate record extending over 15 years. Although MODIS has already exceeded its design lifetime, calibration and validation efforts continue to maintain the calibration accuracy of the instruments to ensure the integrity of the ongoing observations. Studies based on MODIS data provide some of the best examples of the application of space-based AOD to the understanding of global climate and regional air-quality conditions (see the third section below).

Two primary AOD products are produced from the MODIS observations based on the Dark Target[2] and Deep Blue[3] aerosol-retrieval algorithms. These two products are made readily available to users. A third research product from MODIS, Multi-Angle Implementation of Atmospheric Correction (MAIAC), has been used in many studies but is not yet processed operationally.

Dark Target: The NASA Dark Target algorithm produces AOD over ocean and other dark land (for example, vegetated) surfaces using two separate algorithms for the different regimes. The standard product for climate studies is AOD aggregated at 10-kilometer resolution (Levy and others 2013). Over land, the retrieval produces AOD at 550 nanometers, the AOD model/weight, and surface reflectance at 2,130 nanometers based on 500-meter resolution observations at 470, 650, and 2,130 nanometers and using a LUT of RT-model–computed reflectances as a function of aerosol and surface properties. Dark targets are identified as pixels where the 2,130-nanometer reflectance is between 0.01 and 0.25. For these locations, the quality flag is set based on the goodness of the fit between the observed and modeled reflectance. Targets with reflectance up to 0.4 are processed through a separate algorithm path but flagged as reduced quality. To meet the growing interest in AOD for regional studies, the product over land is now also produced at a 3-kilometer resolution (Remer and others 2013). The error estimate of the 10-kilometer AOD based only on good-quality retrievals has been assessed against AERONET measurements to be $\pm(0.05 + 0.15$ $AOD_{AERONET})$,[4] whereas the error in the 3-kilometer product is $\pm(0.05 + 0.20$ $AOD_{AERONET})$. This product is available through the NASA Level-1 and Atmospheric Archive and Distribution System (LAADS) Distributed Active Archive Center (DAAC)[5,6] or through the NASA WORLDVIEW website.[7]

Investigations of the use of the 3-kilometer Dark Target product to provide better definition of local aerosol gradients (Munchak and others 2013) suggest that the 3-kilometer product does provide better spatial coverage (see figure A.2). However, this product tends to overestimate the aerosol loading and can be susceptible to noise problems in urban areas, likely because of inadequate characterization of surface features. For such applications, some caution is advised.

Deep Blue: The Deep Blue aerosol algorithm explicitly models surface reflectance, rather than estimating it from observations at 2,130 nanometers as in the Dark Target algorithm. This explicit modeling allows the Deep Blue algorithm to be applied to bright surfaces such as desert, semiarid, and urban regions (Hsu and others 2013). To enable retrievals over brighter surfaces, the algorithm uses the 412-nanometer or "deep blue" band on MODIS (where the surface reflectance over land is much lower than in low-frequency visible bands), thus providing better contrast with the aerosol signal. This product is computed over land at 1-kilometer resolution and aggregated to 10 kilometers. Surface reflectance is determined by one of several methods depending on the surface type, including use of a database that takes account of seasonal changes and the variability of urban landscapes based on the normalized difference vegetation index (NDVI).

FIGURE A.2

MODIS Dark Target 10-kilometer and 3-kilometer aerosol-optical-depth products retrieved for clear land and ocean fields of view and the local 5-kilometer average derived from the products, outer circle, compared to ground-based measurements, inner circle, over Baltimore, US

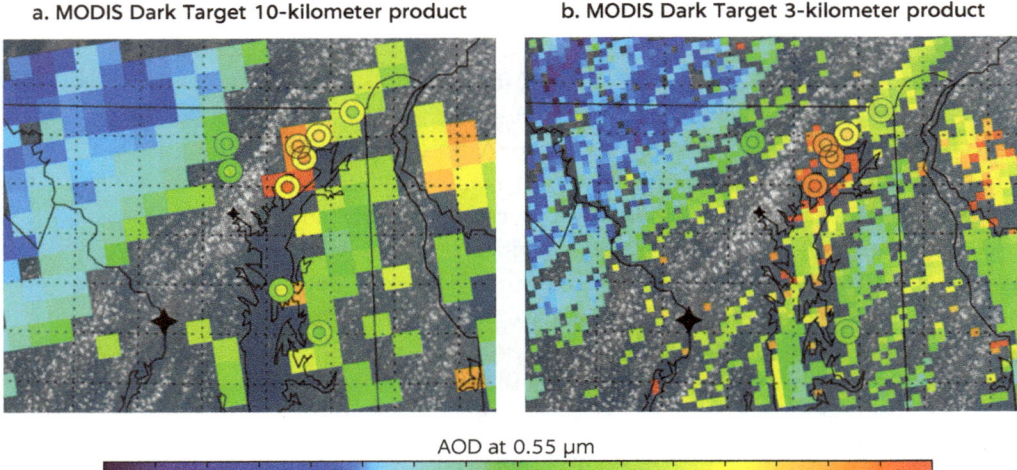

Source: Munchak and others 2013. Note that these results used an older version of the MODIS surface reflectance scheme, and more recent versions (for example, Gupta and others 2016) show much better performance over urban areas.
Note: AOD = aerosol optical depth; MODIS = Moderate-Resolution Imaging Spectroradiometer.

The (prognostic) uncertainty derived by Deep Blue is comparable to the error estimate of the Dark Target product for typical aerosol levels in the range from 0.1 to 0.5, $\pm 0.03 + 0.2\tau_{DB}$.

The combined Deep Blue and Dark Target aerosol product was developed to provide users with the alternative of a single merged MODIS product (Sayer and others 2014). It applies NDVI criteria based on a monthly composite to select either the Dark Target or Deep Blue AOD for a given 10-kilometer pixel location:

- NDVI ≤ 0.2   Use Deep Blue
- NDVI ≥ 0.3   Use Dark Target
- 0.2 < NDVI < 0.3   Use product with highest quality or report mean value.

As a result, Deep Blue is reported over desert regions, Dark Target is selected over permanent vegetation, and in other areas the selection of Deep Blue or Dark Target varies with the season. Comparison of the two products to AERONET indicates that neither Deep Blue nor Dark Target consistently outperforms the other (see map A.1). Dark Target tends to have slightly smaller overall errors compared to AERONET, but Deep Blue provides additional coverage and tends to perform better for low-AOD conditions. Both the Deep Blue and the combined product are available through the NASA LAADS DAAC or through the NASA WORLDVIEW website.

MAIAC: The Multi-Angle Implementation of Atmospheric Correction (MAIAC) algorithm[8] is an atmospheric correction algorithm designed for

MAP A.1

Fraction of good-quality attempted retrievals from Deep Blue and Dark Target algorithms showing differences in coverage over desert regions, and showing differences in coverage due to the quality checks applied in each algorithm

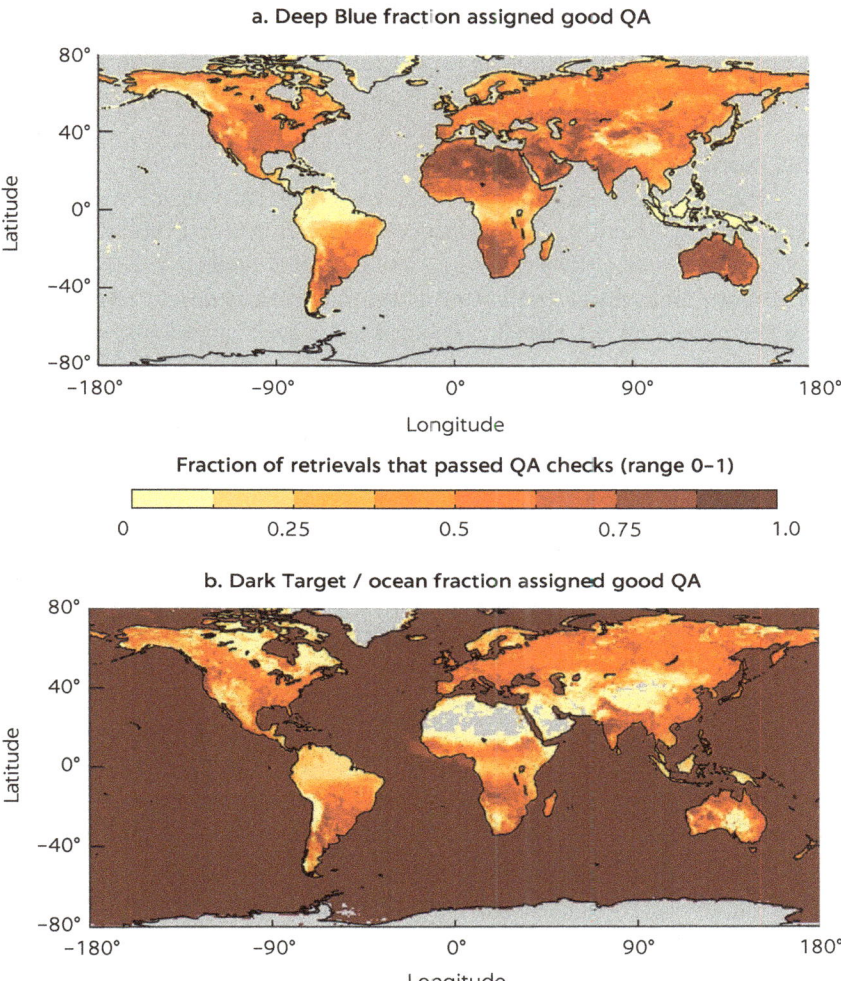

a. Deep Blue fraction assigned good QA

Fraction of retrievals that passed QA checks (range 0–1)

b. Dark Target / ocean fraction assigned good QA

Source: Sayer and others 2014.
Note: QA = quality assurance.

MODIS that performs simultaneous retrievals of atmospheric aerosols and bidirectional surface reflectance (Emili and others 2011; Lyapustin and others 2011). Unlike the Dark Target and Deep Blue algorithms, the MAIAC algorithm does not rely on any predetermined relationships with the shortwave information to constrain the surface retrieval but instead invokes a multiangle, multitemporal retrieval of the surface bidirectional reflectance distribution function and aerosol loading, based on assumptions that the surface reflectance does not change over a 16-day period and that AOD changes little over 25-kilometer distances. The resulting AOD product is generated on a one-kilometer sinusoidal grid. Over vegetated regions (forest, cropland, grassland, and savanna), more than 66 percent of retrievals are within the expected error $\pm(0.05 + 0.05t)$ with a correlation coefficient to AERONET better than 0.86 (Martins and others 2017).

Results over other bright backgrounds were less precise, especially for low aerosol signals where the retrieval problem is not as well constrained (for example, 55 percent of retrievals within the expected error for urban regions).

The MAIAC aerosol product is nominally distributed through the NASA LAADS DAAC[9,10] but appears to be currently unavailable except for a limited data set over the Amazon basin in South America from 2000 to 2012.[11,12] Online information indicates support of the product is ongoing[13,14,15] with a potential extension to process VIIRS data in the works.[16]

VIIRS products

The Visible Infrared Imaging Radiometer Suite (VIIRS) sensor is a scanning radiometer on Suomi-National Polar-orbiting Partnership (Suomi-NPP) and the Joint Polar Satellite System (JPSS) satellites operated by NOAA and NASA. Over land, the NOAA VIIRS AOD product[17] is generated with a LUT-based retrieval of AOD, based on five aerosol models using observed reflectance in 750-meter moderate-resolution bands at 412, 445, 488, 672, and 2,250 nanometers (Jackson and others 2013). The surface reflectance in the red (672 nanometers) and blue (488 nanometers) bands is inferred based on an empirical relationship with the shortwave infrared (2,250 nanometers) band that holds mainly for dark surfaces. The AOD is then determined from observed blue-to-red reflectance ratios by matching the observed signal to computed values for a given aerosol model. AOD retrievals are not performed for clouds, snow, fire, glint, and other bright surfaces identified, based on a shortwave infrared NDVI threshold ($NDVI_{SWIR}$ < 0.05 with a 2,250-nanometer reflectance > 0.3). $NDVI_{SWIR}$ > 0.2, consistent with vegetation backgrounds, is required for high-quality retrievals.

The AOD output is available as an intermediate product at 750-meter resolution and as a 6-kilometer (eight-by-eight aggregated) product, thus providing some advantage over the MODIS 3- and 10-kilometer products for analysis of the localized aerosol distribution. While studies of regional applications of VIIRS AOD products have been presented as conference papers, no comprehensive review of the product has been published. The overall performance of the VIIRS product has been validated against requirements with uncertainties similar to that of MODIS Dark Target (Huang and others 2016). However, comparisons of the 1-kilometer MAIAC AOD product to the high-resolution VIIRS product (see figure A.3) found that the VIIRS product was biased to higher AOD in urban and mountainous regions and was susceptible to errors in regions of snowmelt (Superczynski, Kondragunta, and Lyapustin 2017). The official NOAA VIIRS products are distributed through the Comprehensive Large Array-Data Stewardship System (CLASS)[18] and with the most recent 90 days available via file transport protocol (FTP).[19]

Alternatives to the operational NOAA VIIRS AOD product are in development. For example, the NASA Dark Target algorithm was applied to VIIRS data, and results were compared with the operation NOAA algorithm and to MODIS (Levy and others 2015). The NASA Deep Blue algorithm[20] also has been applied to VIIRS data. In addition, the NOAA Center for Satellite Applications and Research (STAR) has published work describing an enhanced AOD algorithm capable of retrieving aerosols over brighter surfaces and comparable to the Deep Blue product (Zhang and others 2016). Finally, VIIRS products based on the MAIAC algorithm have been referenced in online presentations, but the status of such products is unclear. Once available, these alternative products

FIGURE A.3

Comparison of VIIRS 750-meter and MAIAC one-kilometer aerosol-optical-depth products during western US fires, 2013

a. VIIRS 750-meter AOD product

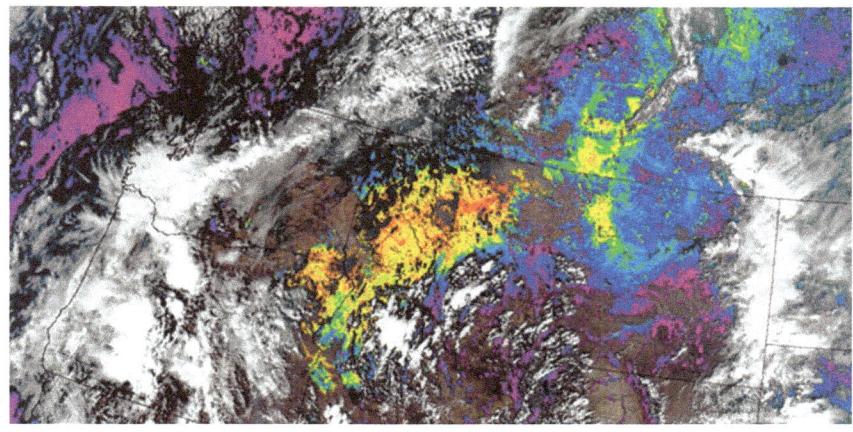

b. MAIAC one-kilometer AOD product

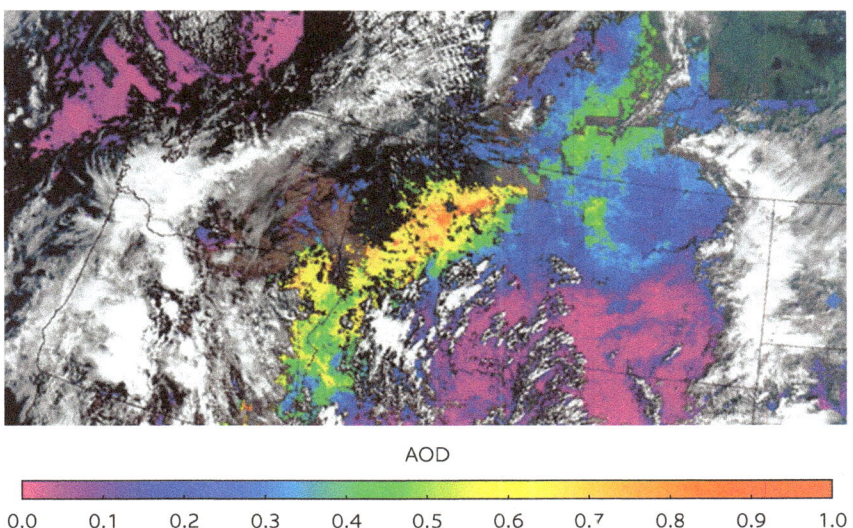

AOD

0.0 0.1 0.2 0.3 0.4 0.5 0.6 0.7 0.8 0.9 1.0

Source: Superczynski, Kondragunta, and Lyapustin 2017.
Note: AOD = aerosol optical depth; MAIAC = Multi-Angle Implementation of Atmospheric Correction; VIIRS = Visible Infrared Imaging Radiometer Suite.

may provide improved coverage, in particular for urban areas, potentially increasing their utility for regional air-quality applications.

MISR products

The Multi-angle Imaging Spectroradiometer (MISR) instrument uses multiangle techniques in the VIS and NIR to acquire stereoscopically resolved observations of AOD under sunlit conditions (Diner and others 1998). All 36 channels (nine cameras and four spectral bands) are used when performing aerosol retrievals over land (Martonchik and others 2004). MISR measurements (between 70.5° forward and 70.5° aft) span a wide range of scattering angles and air mass factors providing information on aerosol microphysical properties and yielding sensitivity to optically thin aerosols. The MISR retrieval approach relies on describing the change in surface contrast with view (camera) angle over a region 17.6 kilometers by 17.6 kilometers in size, composed of 16 by 16 (256) subregions, each 1.1 kilometer by 1.1 kilometer. This subregion size is the nominal spatial resolution of MISR when observing in Global Mode, and 17.6 kilometers is the spatial resolution of MISR's retrieved aerosol product. Aerosol retrievals are performed on only a regional and not a subregional basis, because the entire region is used to determine the surface contribution to the reflectance. In addition to having a lower horizontal resolution than the MODIS and VIIRS products, the MISR instrument has a smaller swath than MODIS (400 kilometers versus 2,330 kilometers) and thus only observes a given location once every nine days, unlike once a day for MODIS and VIIRS.

The performance of AOD retrievals from MISR is comparable to that of MODIS. Martonchik and others (2004) compared MISR AOD with AERONET observations and found that at 17.6-kilometer spatial resolution, the estimated uncertainty is about 0.08. When the spatial resolution was degraded to 52.8 kilometers, the estimated MISR AOD uncertainty decreased to about 0.05. Khan and others (2005) found that 66 percent of the measurements agreed with AERONET within $\pm(0.05 + 0.2\tau)$. Kahn and others (2009) compared MISR with MODIS AOD and found that where coincident AOD retrievals are obtained over the ocean, the MISR-MODIS correlation coefficient is about 0.9 with a slope of 0.75; over land, the correlation coefficient is about 0.7 with a slope of 0.60. MISR AOD data are available from the Jet Propulsion Laboratory.[21]

AVHRR products

The Advanced Very High Resolution Radiometer (AVHRR/3) instrument provides visible-through-IR remote sensing observations on the LEO satellites flown by NOAA (for example, NOAA-15) and by the European Organisation for the Exploitation of Meteorological Satellites (EUMETSAT) (for example, Metop-A). However, the operational NOAA AOD product[22] is produced only over oceans, though the application for AOD retrievals over land has also been investigated (Li and others 2013). The NASA Deep Blue algorithm has been applied to the NOAA data to produce an aggregated 8.8-kilometer product over land and ocean with an expected error over land of $\pm(0.05 + 0.25\tau)$ (Sayer and others 2017). This application of Deep Blue uses a database of surface reflectance for retrievals over bright surfaces, but because AVHRR lacks a 412-nanometer band, the red band at 620 nanometers is used instead. Because the surface at 630 nanometers is not as dark as 412 nanometers, the sensitivity of the aerosol retrieval is limited compared to MODIS and VIIRS. AVHRR also lacks a shortwave band that is used in the MODIS/VIIRS algorithms to derive the surface reflectance for dark targets. For AVHRR, the surface is modeled as an empirical function of NDVI. However, this product was available online only through 2011.[23]

EUMETSAT produces an operational multisensor aerosol product based on observations from AVHRR, the Global Ozone Monitoring Experiment (GOME; a medium-resolution double UV-VIS spectrometer), and the infrared atmospheric sounding interferometer (IASI)–Fourier transform spectrometer.[24,25] However, the primary input for the AOD retrieval is derived from GOME, and the resolution of the product (5 kilometers by 40 kilometers for

Metop-A or 10 kilometers by 40 kilometers for Metop-B) is dictated by this instrument.

Geostationary satellites

GOES-NOP imager products

The GOES-N series geostationary satellites (GOES-NOP; for example, GOES-13 East at 75° W and GOES-15 West at 135° W) operated by NOAA are able to provide Full Disk imagery of the Earth covering North and South America in a single VIS band and four IR bands with a refresh rate of 30 minutes. For over 15 years, NOAA has supported the generation of an operational AOD product based on the GOES observations. Being limited to a single (uncalibrated) VIS band, the accuracy of this product is somewhat less than other, more advanced products, but the GOES product is still of significant value owing to its unique temporal information. The GOES Aerosol/Smoke Product (GASP)[26] is derived using a surface reflectance based on a 28-day composite (converted to albedo using an RT model) and AOD retrieved using an RT-model–computed LUT and comparing observations to model reflectances in the VIS band (Knapp and others 2005). The retrieval is limited to a single continental aerosol model and restricted to regions with dark vegetation. GASP is produced at four-kilometer resolution (at nadir) with a quoted precision of ±0.13 and a correlation of R of 0.72 with AERONET. The current product is limited to the continental United States, and products are available in binary format in NRT from the NOAA Satellite Products and Services Division website.[27] This product was discontinued in 2018 from GOES-13 (East) when the GOES-16 products became operational.

SEVIRI-MSG products

The SEVIRI-MSG Aerosol Over Land (SMAOL) product[28] (based on observations from the Spinning Enhanced Visible and Infrared Imager [SEVIRI] on Meteosat Second Generation [MSG]) was developed by HYGEOS for the Centre National de la Recherché Scientific (CNRS). SMAOL is distributed in NRT though the ICARE Data and Services Center (Bernard and others 2011; Mei and others 2012). The algorithm approach is similar to that of the GOES GASP product but with greater sensitivity. SEVIRI observations at 630, 810, and 1640 nanometers (corrected for gas absorption and molecular scattering) are used to derive AOD at 550 nanometers chosen from between five aerosol models using a LUT approach to minimize the differences at the three wavelengths. The product is generated at the three-kilometer (at nadir) resolution and produced every 15 minutes during daytime/cloud-free conditions and subject to some viewing and scattering angle restrictions. With the MSG satellite at 0° longitude, the product provides coverage of Europe and Africa. The surface reflectance used in the retrieval is based on a fit to the minimum reflectance over a 14-day period. This methodology is valid only for dark targets; therefore, desert regions (such as the Sahara, Sahel, and Namib) are excluded. The product has been validated with comparisons to AERONET and to the MODIS Dark Target AOD product and found to provide good estimates of both diurnal and daily variations in AOD (see figure A.4). Error sources include possible subpixel cloud contamination, errors in surface reflectance estimation, and possible temporal noise in model selection.

FIGURE A.4

Detection of temporal variations in aerosol optical depth with the SEVIRI-MSG aerosol over land product compared to AERONET measurements at Palaiseau, France, and to MODIS, July 14, 2006

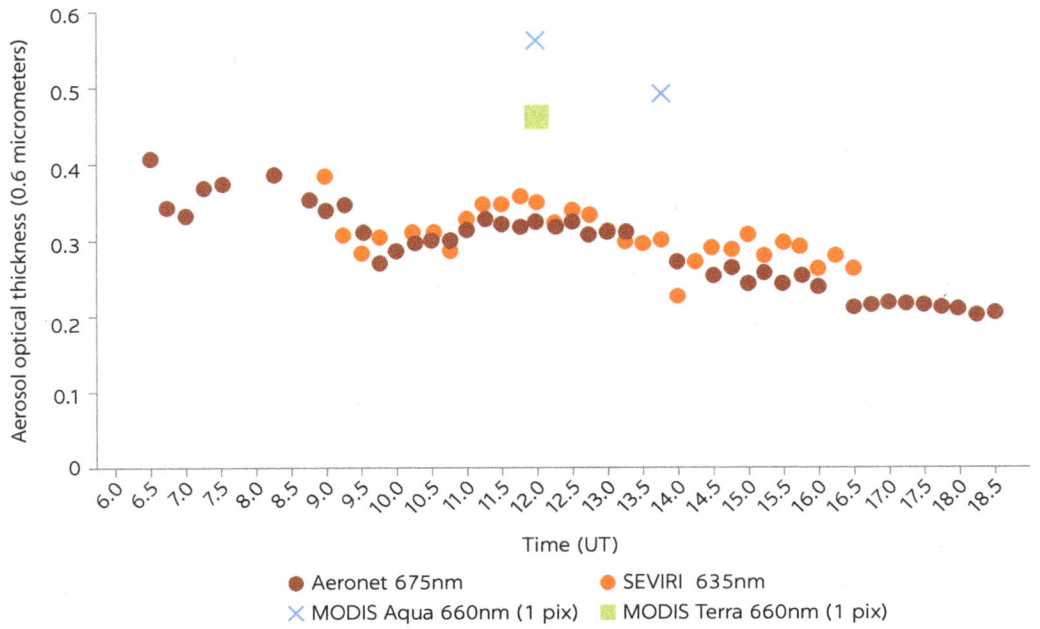

Source: ©Bernard and others 2011; CC BY 3.0.
Note: AERONET = Aerosol Robotic Network; AOT = aerosol optical thickness; MODIS = Moderate-Resolution Imaging Spectroradiometer; MSG = Meteosat Second Generation; nm = nanometer; SEVIRI = Spinning Enhanced Visible and Infrared Imager; UT = Universal Time.

GOES-R ABI products

GOES-R is the first in a series of next-generation geostationary environmental satellites covering the western hemisphere.[29] It was launched as GOES-16 in November 2016. For earth remote sensing, observations are collected by the Advanced Baseline Imager (ABI), which offers much-improved spatial, temporal, and spectral information over the preceding GOES-NOP Imager series. For the past year, the ABI instrument products have undergone an intensive validation process. In December 2017, the satellite was moved to the GOES East position, with products scheduled to become operational in 2018.

The GOES-R AOD and aerosol-particle–sized products[30,31] are derived from ABI reflectance measurements through physical retrievals that utilize a LUT of TOA reflectance that is precalculated using an RT model. Retrievals are performed separately over land (for dark surfaces) and ocean. The baseline AOD product is generated at two-kilometer resolution (at nadir) for Full Disk and continental United States (CONUS) geographic coverage areas. The Full Disk AOD output product is generated at 15-minute intervals (see figure A.5), whereas the CONUS product is generated at five-minute intervals. The GOES-R product performance specification is a function of AOD value. For intermediate AOD loading between 0.04 and 0.8, the accuracy over land is ± 0.04 and the precision

FIGURE A.5

Full disk coverage at 550 nanometers of GOES-R SMAOD product over ocean and land

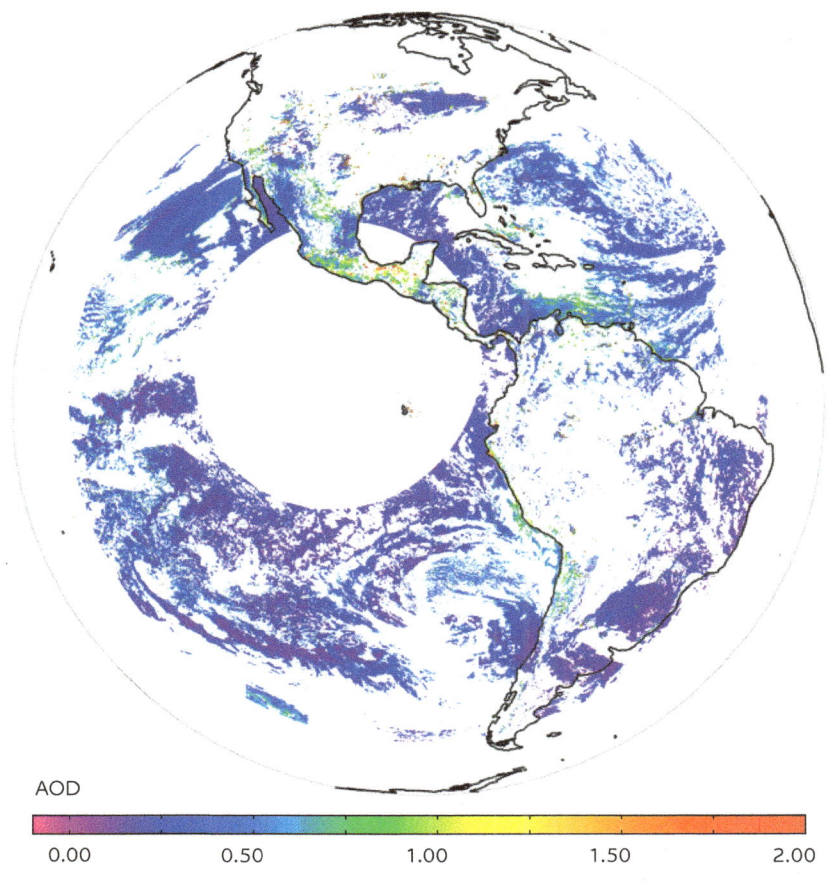

AOD

| 0.00 | 0.50 | 1.00 | 1.50 | 2.00 |

Source: World Bank.
Note: AOD = aerosol optical depth; GOES = Geostationary Operational Environmental Satellite R series; SMAOD = suspended matter, aerosol optical depth.

is ± 0.25, which is similar to the equivalent polar products. Once approved, the GOES-R AOD product will be distributed through the NOAA's Comprehensive Large Array-Data Stewardship System.[32,33]

Although the GOES-R program product is the official operational AOD product, alternative algorithms have been developed to also measure aerosol properties from GOES-R data. For example, the Enterprise Processing System (EPS) AOD algorithm[34] is an algorithm from NOAA STAR intended to support the generation of an AOD product from GOES-R ABI and JPSS VIIRS with a common methodology.[35] The status of the NOAA EPS product is uncertain at this time.

Himawari-8/9 AHI products

Himawari-8/9 is a new series of geostationary weather satellites operated by the JMA. The imagery collected by Himawari-8/9 is produced by the Advanced Himawari Imager (AHI), which is very similar in design to the ABI instrument flown on GOES-16. Although the MODIS Deep Blue, GOES-R, and EPS algorithms have been applied and tested on AHI data, these products are not currently available. The JMA advertises their own environmental products,[36,37] but these are also not freely available.[38,39]

Case studies

In table A.2 three urban locations with poor air quality are reviewed to evaluate the available satellite AOD products and to assess the utility and limitations of these products for the regions. Each location includes a link to the NASA WORLDVIEW website with visualizations of the Dark Target and Deep Blue aerosol products from MODIS-Terra and MODIS-Aqua.

In Delhi, India, the air quality in late fall is affected by widespread smoke from rural crop fires combined with the localized urban smog trapped in the

TABLE A.2 Case study of satellite aerosol-optical-depth applicability for three representative low- and middle-income country urban areas

LOCATION	COORDINATES	LAND TYPE/CLIMATE	COVERAGE
Delhi, India	28°36′36″N, 77°13′48″E	Land type: Semivegetated, desert, urban	MODIS
		Topography: Yamuna River basin	VIIRS
		Climate: Rain Jun to Oct	MISR
		Air quality: Widespread smoke from rural crop fires mixed with urban smog	WORLDVIEW[a]
Lima, Peru	12°2′36″S, 77°1′42″W	Land type: Coastal desert, urban, semivegetated	MODIS
		Topography: Andes foothills	VIIRS
		Climate: Persistent clouds/fog May to Nov	MISR
		Air quality: Urban (localized) smog	ABI
			WORLDVIEW[b]
Ulaanbaatar, Mongolia	47°55′N, 106°55′E	Land type: Grasslands (steppes)	MODIS
		Topography: Tuul River valley	VIIRS
		Climate: Seasonal snow cover	MISR
		Air quality: Desert dust, urban smog, coal/wood fires (winter)	AHI
			WORLDVIEW[c]

Source: World Bank, produced with Esri ArcGIS.

Note: AHI = Advanced Himawari Imager; MISR = Multi-angle Imaging Spectroradiometer; MODIS = Moderate-Resolution Imaging Spectroradiometer; VIIRS = Visible Infrared Imaging Radiometer Suite.

a. https://worldview.earthdata.nasa.gov/?p=geographic&l=MODIS_Aqua_SurfaceReflectance_Bands143(hidden),MODIS_Terra_SurfaceReflectance_Bands143(hidden),MODIS_Aqua_CorrectedReflectance_TrueColor(hidden),MODIS_Terra_CorrectedReflectance_TrueColor,MODIS_Terra_NDVI_8Day(hidden),MODIS_Terra_AOD_Deep_Blue_Combined(hidden),MODIS_Terra_AOD_Deep_Blue_Land(hidden),MODIS_Terra_Aerosol(hidden),MODIS_Terra_Aerosol_Optical_Depth_3km(hidden),MODIS_Aqua_Aerosol(hidden),MODIS_Aqua_AOD_Deep_Blue_Combined(hidden),MODIS_Aqua_Aerosol_Optical_Depth_3km(hidden),MODIS_Aqua_AOD_Deep_Blue_Land(hidden),Reference_Labels,Reference_Features(hidden),Coastlines&t=2017-03-04&z=3&v=75.58614905641733,27.66425081916523,78.66232093141733,29.24825472541523&ab=off&as=2017-03-04&ae=2017-03-11&av=3&al=true.

b. https://worldview.earthdata.nasa.gov/?p=geographic&l=MODIS_Aqua_SurfaceReflectance_Bands143(hidden),MODIS_Terra_SurfaceReflectance_Bands143(hidden),MODIS_Aqua_CorrectedReflectance_TrueColor(hidden),MODIS_Terra_CorrectedReflectance_TrueColor,MODIS_Terra_NDVI_8Day(hidden),MODIS_Terra_AOD_Deep_Blue_Combined(hidden),MODIS_Terra_AOD_Deep_Blue_Land(hidden),MODIS_Terra_Aerosol(hidden),MODIS_Terra_Aerosol_Optical_Depth_3km(hidden),MODIS_Aqua_Aerosol(hidden),MODIS_Aqua_AOD_Deep_Blue_Combined(hidden),MODIS_Aqua_Aerosol_Optical_Depth_3km,MODIS_Aqua_AOD_Deep_Blue_Land(hidden),Reference_Labels,Reference_Features(hidden),Coastlines&t=2017-05-25&z=3&v=-77.91445867492484,-12.577473819870836,-76.37857000304984,-11.608479679245836&ab=off&as=2017-03-04&ae=2017-03-11&av=3&al=true.

c. https://worldview.earthdata.nasa.gov/?p=geographic&l=MODIS_Aqua_SurfaceReflectance_Bands143(hidden),MODIS_Terra_SurfaceReflectance_Bands143(hidden),MODIS_Aqua_CorrectedReflectance_TrueColor(hidden),MODIS_Terra_CorrectedReflectance_TrueColor,MODIS_Terra_NDVI_8Day(hidden),MODIS_Terra_AOD_Deep_Blue_Combined(hidden),MODIS_Terra_AOD_Deep_Blue_Land(hidden),MODIS_Terra_Aerosol(hidden),MODIS_Terra_Aerosol_Optical_Depth_3km(hidden),MODIS_Aqua_Aerosol(hidden),MODIS_Aqua_AOD_Deep_Blue_Combined(hidden),MODIS_Aqua_Aerosol_Optical_Depth_3km(hidden),MODIS_Aqua_AOD_Deep_Blue_Land(hidden),Reference_Labels,Reference_Features(hidden),Coastlines&t=2017-04-25&z=3&v=106.0165808187288,47.38272533863788,107.5524694906038,48.35171947926288&ab=off&as=2017-03-04&ae=2017-03-11&av=3&al=true.

Yamuna River basin around the city. When conditions are clear, the MODIS Deep Blue product typically provides complete coverage across the region, but the Dark Target product is excluded from parts of the urban area where the background signal is high. Also, the coverage from the Dark Target algorithm varies with the season depending on the stage of vegetation growth in the surrounding rural area. On very bad days the smoke can be so widespread and thick that localized variations in air quality cannot be determined. During these periods, the Dark Target product is sometimes not produced at all for the region.

The pollution in Lima, Peru, is largely due to localized urban smog. Lima is located on the west coast of South America with weather patterns greatly influenced by the Humboldt Current and its location just west of the foothills of the Andes mountains. For significant portions of the year (May to November), Lima is often engulfed in fog, and during these periods the remote sensing of AOD is not possible. In addition, the background signal from the city and the surrounding desert is relatively bright, such that the coverage from the Dark Target product is minimal. The Deep Blue product does provide coverage over the urban environment but with frequent gaps due to clouds. In addition, as Lima is situated on the coast, AOD retrievals are not produced for any pixels with a mix of land and water, further reducing the coverage of the Deep Blue product for this city.

Ulaanbaatar, Mongolia, is in the Tuul River valley at the foot of the heavily forested Bogd Kahn Uul mountains and is surrounded by a steppe ecoregion where the largely grassland vegetation varies with the seasons. In winters, the entire region is covered with snow. The air-quality problems in Ulaanbaatar are highest in the winter months when use of coal and wood for heat gives rise to significant smoke emissions that couple with the local smog from cars and other sources. Because satellite-based AOD retrievals are not capable of distinguishing aerosol from the bright background of snow, neither the Dark Target nor Deep Blue products are useful under these conditions. At other times of year on clear days, the Deep Blue product does provide fairly good coverage over the region, but the coverage from the Dark Target algorithm is spotty, often limited only to the nearby vegetated mountainous areas and not providing information in the urban and steppe environments.

CONVERTING AOD TO GROUND-LEVEL PM$_{2.5}$

Many studies have attempted to convert satellite AOD to ground-level PM$_{2.5}$ concentration estimates using either statistical techniques (the first subsection below), approaches based on the chemical transport model (CTM) (discussed in the second subsection), or hybrid approaches that mix the two (discussed in the third subsection).

Statistical approaches

Several studies have used a purely statistical approach, where linear mixed effects models (for example, Hu, Waller, Lyapustin, Wang, Al-Hamdan, and others 2014; Sorek-Hamer and others 2015) or nonlinear generalized additive models (GAMs; Sorek-Hamer and others 2013; Strawa and others 2013) have been trained on historical GLM data to predict ground-level PM$_{2.5}$ using AOD and other meteorological and geographic variables as input variables. The advantages of statistical approaches are that the statistical models can be trained for specific areas and are easier to run than CTMs. However, unlike CTM approaches, statistical models require a substantial amount of GLM data for the model training and can be used only in the region in which they were trained.

Statistical approaches also provide an estimate of the uncertainty in their predictions. In addition, the uncertainty in AOD, planetary boundary layer (PBL) height, and other variables can be propagated to the ground-level $PM_{2.5}$ estimates by running the statistical model for the high and low error bounds of the variable, and then combining that uncertainty with the estimated uncertainty of the statistical fit itself.

Linear mixed-effects models

Linear mixed-effects models assume a linear relationship between the predictors and the modeled variable but allow the intercepts and slopes to vary with the day by fitting fixed (that is, constant) and random (that is, daily varying) values for those parameters, with the limitation that the random values cannot also vary with location. Routines for training these models are included in the open source R statistical program.

Hu, Waller, Lyapustin, Wang, Al-Hamdan, and others (2014) used a two-stage linear mixed-effects model to estimate ground-level $PM_{2.5}$ concentrations at one-kilometer resolution from MAIAC AOD data over the Southeast United States. In the first stage the predictors included MAIAC AOD, wind speed, and elevation, as well as the length of major roads, percent forest cover, and point emissions of $PM_{2.5}$ within one kilometer of the site. The second stage used geographically weighted regression (GWR) to predict the residuals from the first-stage model using the MAIAC AOD. GWR is an extension of least-squares regression that allows predictor coefficients to vary spatially by weighting the estimate-observation pairs according to the inverse-squared distance from individual observation sites, resulting in a spatially continuous prediction of $PM_{2.5}$ over an area at one-kilometer resolution.

The GLM data used in the training had 166 monitors over a domain of 800 by 1,200 square kilometers. Hu, Waller, Lyapustin, Wang, Al-Hamdan, and others (2014) found that all of the first-stage predictors were significant at an $\alpha = 0.05$ level. The first-stage model explained 64 percent of the variability with a mean error of 2.8 micrograms per cubic meter and a root-mean-square error (RMSE) of 3.9 micrograms per cubic meter. The second-stage GWR model increased the variability explained to 67 percent and reduced the mean error to 2.5 micrograms per cubic meter but had little impact on the RMSE.

Sorek-Hamer and others (2015) used a linear mixed-effects model to predict ground-level $PM_{2.5}$ at 10-kilometer resolution over Israel, using MODIS Deep Blue AOD data to provide better retrievals over the desert surfaces in the area. The MODIS AOD was the only predictor variable used in the model. Over Israel, this model explained 45 percent of the variability in $PM_{2.5}$ and 69 percent of the variability in PM_{10}, with RMSE of 12.1 and 27.9 micrograms per cubic meter, respectively.

Lv and others (2016) developed a method to account for the spatial and temporal variations in $PM_{2.5}$ and missing AOD observations before applying a Bayesian hierarchical mixed-effects model. This gap-filling approach uses the observed seasonal mean $AOD/PM_{2.5}$ ratio at a given site and the measured $PM_{2.5}$ concentration to produce an estimated AOD for the site even on days when clouds or other effects prevent AOD retrieval. Ordinary kriging (see the section "Co-kriging" in this appendix) is then applied to the retrieved and estimate AOD values to provide a daily AOD field over the entire region. Temperature, relative humidity, and planetary boundary layer height (PBLH) were taken from NCEP forecast data, and MODIS land cover and USGS elevation data were also used as spatial predictors. The final model explained 78 percent of the variability in $PM_{2.5}$ in north China and provided ground-level $PM_{2.5}$ concentration fields at 12-kilometer resolution with complete spatial coverage.

Generalized additive models

Generalized additive models (GAMs) are a generalization of linear regression models that are able to account for the potentially nonlinear dependence of the modeled variable on the values of the predictors. The functional dependence of each predictor is determined during the fit as a linear combination of basis functions, with a penalty applied for the number of degrees of freedom included in each functional form. Routines for training GAMs are included in the open source R statistical program.

Strawa and others (2013) used a weighted GAM to predict daily $PM_{2.5}$ concentrations at sites in the San Joaquin Valley in California. They found that a weighted GAM including MODIS dark-target AOD, ozone-monitoring instrument (OMI) AOD, OMI tropospheric nitrogen dioxide (NO_2) columns, and a day-of-year variable explained 74 percent of the variability in $PM_{2.5}$, as compared to 17 percent from a linear model.

Sorek-Hamer and others (2013) also used MODIS and OMI data to predict daily $PM_{2.5}$ concentrations in the San Joaquin Valley. Their GAM used MODIS Dark Target AOD, MODIS Deep Blue AOD, OMI tropospheric NO_2 columns, and a day-of-year variable. This explained 61 percent of the variability in $PM_{2.5}$ with an RMSE of 13.0 micrograms per cubic meter.

Sources of planetary boundary layer height data

A key property in determining the relationship between total column AOD and ground-level $PM_{2.5}$ is the height of the planetary boundary layer (PBL; for example, Alexeeff and others 2015; Chatfield and others 2017; Lee, Chatfield, and Strawa 2016), which varies significantly with time of day, geography, season, and meteorological conditions. For example, PBLH can vary on an urban scale with distance from the city center, distance from coastlines, and other factors, as shown in figure A.6. Including predicted PBLH from meteorological model simulations (especially as an AOD/PBLH ratio) has been shown to significantly increase the performance of the statistical approaches (for example, Chatfield and others 2017).

When there is a well-mixed convective PBL and the aerosols in the PBL dominate the total AOD, there should be a nearly linear relationship between ground-level $PM_{2.5}$ and the AOD/PBLH ratio, and thus for PBL variability to account for much of the variability in the relationship of AOD to $PM_{2.5}$. However, for areas with very shallow PBLs, or when a smoke or dust plume is present above the PBL, the total column AOD may have no relationship to the ground-level $PM_{2.5}$ at all. Thus, in those circumstances, the PBLH along with other parameters can be used to filter suspect estimates of ground-level $PM_{2.5}$ concentrations.

Public sources of PBLH data include *in situ* radiosonde and aircraft vertical profiles, ground-based and satellite remote sensing retrievals, and numerical weather prediction model data. These sources are discussed further below.

Radiosonde and aircraft profiles: Various methods are available for calculating the PBLH from radiosonde profiles of temperature, relative humidity, and

FIGURE A.6

Planetary boundary layer height in areas surrounding Washington, DC, US, on July 14, 2011, at 21:00 UTC

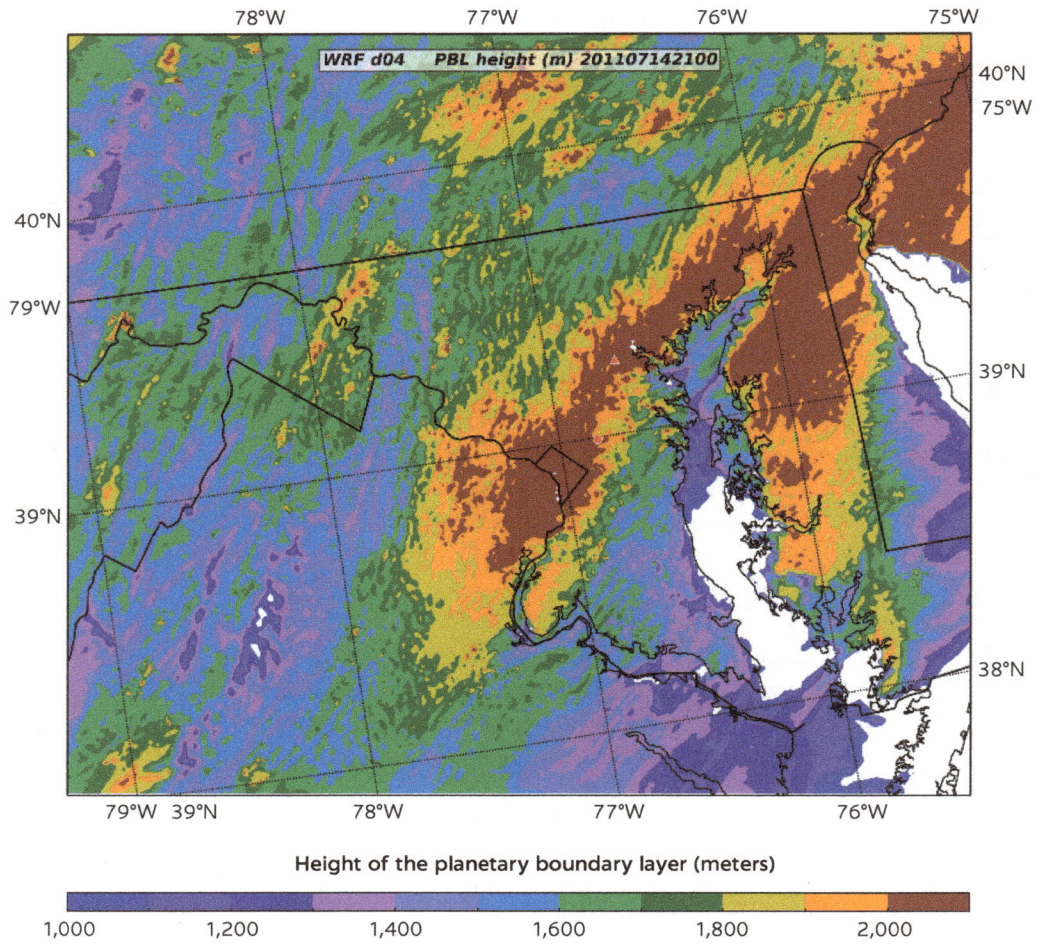

Height of the planetary boundary layer (meters)

Source: World Bank, using Python software.
Note: Planetary-boundary-layer (PBL) height was calculated using the Weather Research and Forecasting (WRF) model.
UTC = Coordinated Universal Time.

wind (for example, Seidel, Ao, and Li 2010). Though considered the gold standard for observing the vertical structure of the atmosphere, radiosondes in global operational networks are generally launched only twice daily and often not at ideal times to estimate the peak daytime PBLH. Furthermore, as shown in map A.2, the radiosonde launch locations are concentrated in higher-income countries in Europe, North America, and Southeast Asia, and there is notably poorer coverage in Africa, Central Asia, and Central and South America. Therefore, the measured profiles and PBLH may not be representative of the cities in LMICs.

The Aircraft Meteorological Data Relay (AMDAR), which includes data from US aircraft from the Meteorological Data Collection and Reporting System (MDCRS), is a global data set produced by commercial aircraft equipped with instruments to measure meteorological data during flights (for example, Drüe and others 2008; Fleming 1996; Zhu and others 2015). Meteorological data

MAP A.2

Radiosonde launch locations for 00:00 UTC, December 8, 2017

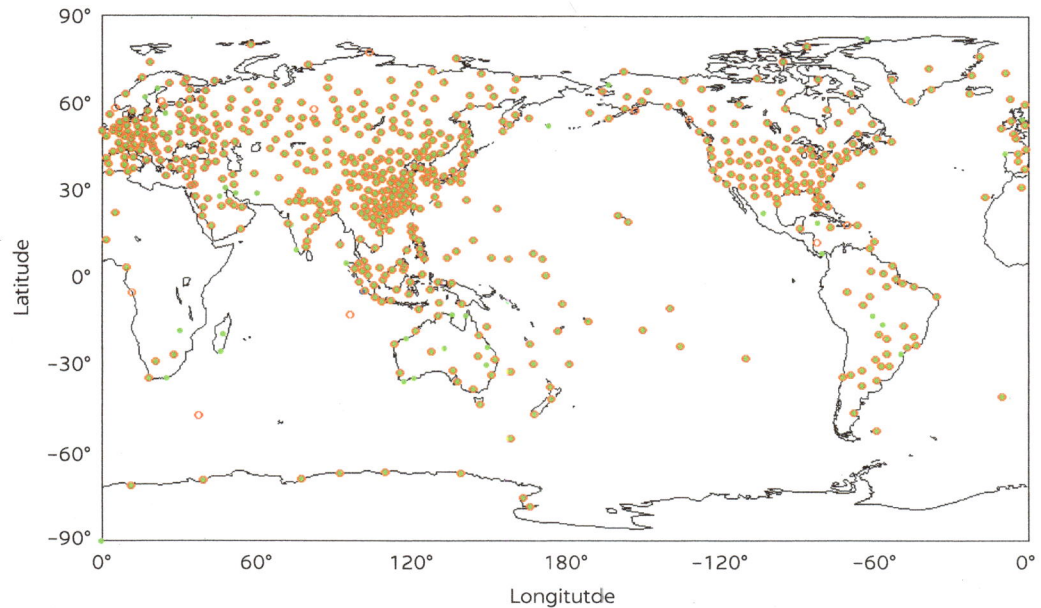

Source: The plot was generated from an application on the Naval Research Laboratory (NRL) website: http://www
.usgodae.org/cgi-bin/cvrg_con.cgi.
Note: UTC = Coordinated Universal Time.

measured at different heights during takeoff and landing have been used to calculate PBLH around the globe (McGrath-Spangler and Denning 2012, 2013).
The AMDAR coverage is more variable than that of radiosondes but also tends
to be concentrated in the higher income regions (map A.3).

Ground-based remote sensing: Ground-based remote sensing systems,
including lidar, ceilometer, sonic detection and ranging (SODAR), and Doppler
wind profilers (DWPs), are capable of providing observations of the PBLH for
field campaigns or as part of operational networks. Of these, currently only
lidar networks routinely produce PBLH products available for public use.
There are three networks contributing to WMO's Global Atmospheric Watch
(GAW) Aerosol Lidar Observations Network (GALION)[40] that provide PBLH
data:

- The European Aerosol Research Lidar Network (EARLINET)[41] consists of
  28 stations distributed over Europe.
- The Asian Dust and Aerosol Lidar Observation Network (AD-Net) includes
  20 stations in Asia including Ulaanbaatar, Mongolia, and Phimai, Thailand,
  with NRT coverage (map A.3).
- The NASA Micro-Pulse Lidar Network (MPLNET) is a federated network of
  currently 23 active Micro-Pulse Lidar (MPL) systems located around the
  globe, including stations at Kanpur, India; Omkoi, Thailand; and Windpoort,
  Namibia (map A.3). An improved PBLH retrieval algorithm has recently been
  incorporated into the latest version (version 3) of the operational product that
  is less susceptible to contamination by clouds and residual layers that can
  result in errors (Lewis and others 2013). Retrievals of PBLH with this new
  algorithm have been validated with PBLH calculated from ozonesonde

MAP A.3

Aircraft observation coverage from AMDAR and MDCRS, December 8, 2017

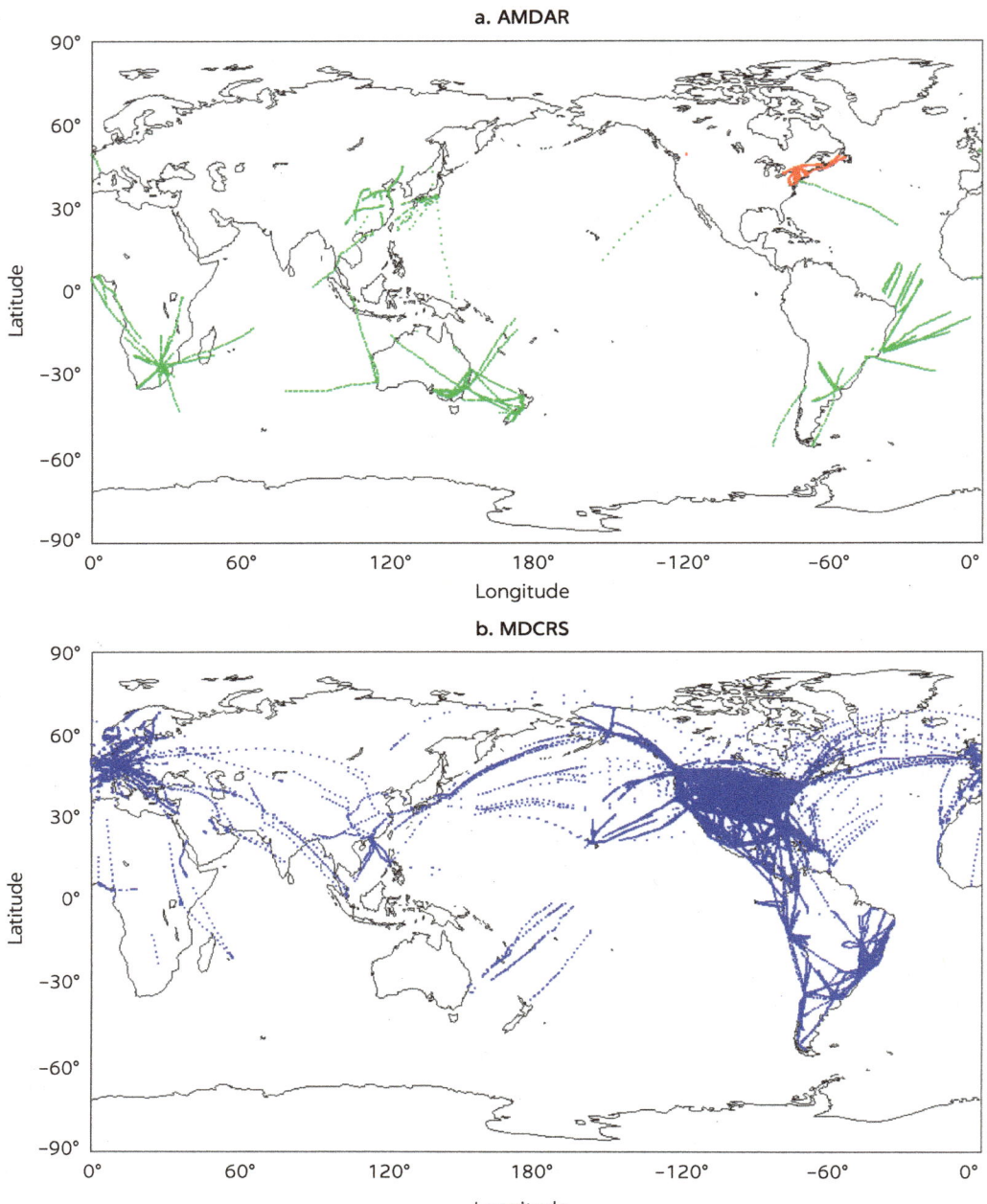

Source: Generated from an application on the Naval Research Laboratory (NRL) website: http://www.usgodae. org/cgi-bin/cvrg_con.cgi.
Note: AMDAR = Aircraft Meteorological Data Relay; MDCRS = Meteorological Data Collection and Reporting System. Red lines = Canadian AMDAR data; green lines = other AMDAR data; blue lines = MDCRS data.

soundings produced during the DISCOVER-AQ field campaign and compared to high-resolution Weather Research and Forecasting (WRF) model simulations (Hegarty and others 2018; Lewis and others 2013). The data are available in NRT (approximately one-hour processing delay).

Satellite data: Only a few studies have examined PBLH using satellite data (Martins and others 2010). Radio occultation data from global positioning system (GPS) satellites have been used to determine PBLH but because of the long tangential path of the GPS signal through the atmosphere the horizontal resolution is coarse ranging from tens to hundreds of kilometers (Ao and others 2012; Guo and others 2011). PBLH retrievals from the Cloud-Aerosol Lidar with Orthogonal Polarization (CALIOP; Winker, Hunt, and McGill 2007; Winker and others 2009) onboard the Cloud-Aerosol Lidar and Infrared Pathfinder Satellite Observations (CALIPSO) satellite have been used to evaluate numerical weather prediction model reanalysis data (Jordan, Hoff, and Bacmeister 2010) during 2006. More recently, a global data set of CALIPSO PBLH retrievals was generated for June 2006 to December 2012, evaluated with PBLHs calculated with AMDAR meteorological data, and used to examine global seasonal variability in the midday PBLH (McGrath-Spangler and Denning 2012, 2013). In addition, CALIPSO PBLHs were compared favorably with radiosondes over China during 2011 to 2014 (Zhang and others 2016). The CALIPSO cloud and aerosol products used to derive the PBLHs for these studies are available through the NASA Langley Data Archive Center (DARC).[42] Unfortunately, the PBLHs were produced independently for only these limited research projects and currently are not yet produced regularly for public dissemination. However, no continuous data sets of satellite-derived PBLHs are publicly available.

NWP model data: Numerical weather prediction (NWP) models assimilate all types of meteorological and environmental observations to produce three-dimensional meteorological fields that are as accurate a representation of the true atmospheric state as possible at a given time. There are two types of NWP data: operational and reanalysis. Operational forecast centers run NWP models several times per day to provide numerical guidance to human weather forecasters and inputs to air-quality forecast models. Operational NWP data have the advantage of being available in near real-time, often within three to four hours of the beginning of a forecast cycle, whereas reanalysis NWP data have a latency period of several months. However, operational NWP models are continuously being updated to correct bugs and address problems in physical parameterization schemes, complicating long-term historical analysis. On the other hand, reanalysis NWP data are generated with models similar to those used in operational centers but frozen in development. When enough changes in the state of NWP modeling accumulates, a new version of the reanalysis, usually with the same starting date as the previous version, is generated with an updated model. Thus, reanalysis data are more appropriate for looking at temporal trends and are also subject to more thorough evaluations and analysis. Both types of NWP data could be used depending on the time requirements on data availability of a particular research task. The observational data described above when available will be used to complement the NWP data, particularly for situations in which the observational data provide a better representation of the local conditions than the grid-averaged NWP data.

Operational NWP: The National Centers for Environmental Prediction (NCEP) Global Forecast System (GFS) is a global model with four daily run cycles beginning at 0000, 0600, 1200, and 1800 UTC (Coordinated Universal Time). For each cycle the GFS model is run for 384 hours, and the initial analysis and forecast fields are available on global latitude-longitude grids of 0.25°, 0.5°, and 1.0° resolution in grib2 format.[43] Forecast output is available at each hour for the first 180 forecast hours, every 3 hours from 180 to 240 hours, and then every 12 hours until 384 hours. The analysis and forecast fields include PBLH and temperature, horizontal winds, vertical velocity, specific and relative humidity, geopotential height, and cloud water mixing ratios at the surface and 30 pressure levels with 25 hecto Pascals (about 200 meters) spacing up to 900 hecto Pascals. The model output is available from the NOAA NCEP ftp server generally within 5 hours after the beginning of the forecast cycle.

The Canadian Meteorological Center (CMC) produces a global NWP forecast twice daily at 0000 and 1200 UTC for 240 hours called the Global Deterministic Prediction System (GDPS).[44] The CMC GDPS initial analysis and forecast data are available on a 0.24° latitude-longitude grid in grib2 format.[45] The data are available every 3 hours out to 140 hours and every 6 hours to 240 hours. The variables include temperature, winds, and relative humidity at the surface and 23 pressure levels. A PBLH diagnostic is not included but could be calculated from vertical model profiles; however, the vertical resolution is coarse ranging from 15 hecto Pascals (about 120 meters) from the surface to 50 hecto Pascals (more than 450 meters) above 900 hecto Pascals, and this would affect the accuracy of the PBLH calculation. The latency time between the beginning of the model cycle and model data availability is approximately 4 to 5 hours.

The JMA runs a Global Spectral Model (GSM) four times a day. The runs at 0000, 0600, and 1800 UTC are for 84 hours, and the 1200 UTC run is for 264 hours. Output data are available in grib2 format on a 0.5° latitude-longitude grid at 3-hour intervals up to 84 hours and then 6-hour intervals up to 264 hours from the WMO Global Information System Centres (GISCs) website. The PBLH diagnostic is not included in the available output fields but could be calculated from temperature, wind, and moisture variables. The data latency time and specific access procedures were not clearly documented on the GISC website. The JMA also produces a mesoscale model forecast with five-kilometer resolution for a domain centered over Japan but including eastern Mongolia. However, only graphical outputs seem to be available in the public domain.

The Fleet Numerical Oceanographic Center produces a global NWP forecast with the Navy Global Environmental Model (Hogan and others 2014) four times per day with a horizontal resolution of about 37 kilometers. However, the gridded output fields are not readily accessible and may be restricted to organizations within the United States.

The European Center for Medium Range Weather Forecasts (ECMWF) and United Kingdom Meteorological Office (UKMET) also produce global NWP forecasts,[46] but the output data are only provided for a licensing fee based on the amount and type of data being requested.

Operational models such as the NCEP GFS are continuously being updated with new physical parameterization schemes (for example, Han and Pan 2011) and new sources of observational data to be assimilated through observing system experiments and observing system simulation experiments (for example, Atlaskin and Vihma 2012). The improvements are often evaluated at the continental to hemispheric scale using geopotential height anomaly correlations and at regional and local scales against temperature, winds, and moisture observations (for example, Cucurull and Derber 2008) but not PBLH. A Google Scholar search for evaluations of operational model PBLH outputs over land produced no results, perhaps because of the general lack of continuous PBLH observations in most locations. Nevertheless, data from operational models are used in satellite algorithms; for example, the MODIS cloud height algorithm uses GFS temperature profile data (for example, Holtz and others 2008). Thus, given the general lack of observations, it seems to be a reasonable procedure to use GFS PBLHs as an estimate of the PBLH in LMICs to determine $PM_{2.5}$ from column AOD measurements. Furthermore, in 2016 NOAA announced the development of a new model to replace the current GFS. The model will still be called the GFS and but will feature a new dynamic core, the Finite Volume on a Cubed Sphere (FV3), that is expected to increase the model's accuracy and numerical efficiency.[47] The new GFS model is projected to be operational in 2019 (Schneider 2016).

MERRA reanalysis: The Modern Era Retrospective-analysis for Research and Applications (Bosilovich 2008; MERRA) is a NASA reanalysis from 1979 to the present day produced using the Goddard Earth Observing System (GEOS) Data Assimilation System (DAS) Version 5 (GEOS-5 DAS; Rienecker and others 2008). The reanalysis data are on the GEOS-5 native 576 by 361 grid with 0.625° by 0.5° resolution at the surface and include PBLH and temperature, winds, relative humidity at the surface, and either 42 pressure levels at three hourly intervals or 72 pressure levels at six-hour intervals. The 42-level product has 8 levels below 800 hecto Pascals, while the 72-level product has 12. An updated version of the reanalysis (Bosilovich and others 2015; Gelaro and others 2017; MERRA-2) extends from 1980 to the present with a latency time of approximately one to two months.

MERRA PBLH data have been evaluated using PBLH retrievals from the CALIPSO satellite (Winker, Hunt, and McGill 2007; Winker and others 2009) over a western hemisphere domain from 60° S to 60° N for August and December 2006 and Africa in August 2006 (Jordan, Hoff, and Bacmeister 2010). The evaluation indicated a better MERRA-CALIPSO correlation in August in both domains (R of 0.73) than in December (R of 0.47). Over the Sahara Desert the MERRA PBLHs were clustered around 1 to 3 kilometers above ground level and generally lower than those of CALIPSO that had several clusters around 1 to 2 kilometers, 3.5 to 4.5 kilometers, and 5 to 6 kilometers above ground level.

The PBLHs from the atmospheric modeling component of GEOS-5 (Atmospheric General Circulation Model) used to derive MERRA have also been evaluated with micro-MPL retrievals at the NASA Goddard Space Flight Center (GSFC) in Maryland for the period of 2001 to 2008 (Lewis and others 2013). The model and MPL diurnal cycles agreed well but the model underestimated the maximum daily PBLH compared to the MPL retrievals by about 0.4 kilometers.

There is inherent uncertainty in any source of PBLH data. This uncertainty can be attributed to three sources: (1) the methods used to calculate it from *in situ* observations (for example, Hegarty and others 2018; Seidel, Ao, and Li 2010), (2) the retrieval method (for example, Lewis and others 2013; Winker, Hunt, and McGill 2007), or (3) error in the NWP model (for example Jordan, Hoff, and Bacmeister 2010; Lewis and others 2013; McGrath-Spangler and Denning 2012, 2013). This uncertainty can be near to 50 percent (for example, Lewis and others 2013; Seidel and others 2012). The uncertainty can also be dependent on the synoptic weather conditions. Hegarty and others (2018) found poor agreement between PBLHs calculated from ozonesondes with different methods, MPL retrievals, and mesoscale model outputs on days with southerly and southwesterly flow in the Baltimore–Washington, DC, area, but good agreement between PBLHs calculated with all methods, MPL retrievals, and model data on days with northerly winds. This suggests that, whenever possible, data from both observations and models should be compared to quantify the uncertainty of the PBLH inputs to the algorithms used to determine the ground-level $PM_{2.5}$ concentrations for satellite AOD retrievals.

Chemical transport model–based approaches

Several studies (for example, van Donkelaar, Martin, and Park 2006) have used CTMs to determine a time-varying relationship between ground-level $PM_{2.5}$ concentrations and satellite AOD observations. This relationship is then used to scale the CTM aerosol profile to match the satellite-observed AOD, providing an improved estimate of the ground-level $PM_{2.5}$ concentration than would be possible from the CTM alone. The advantages of this approach include that it can be generalized to apply to any region of the globe, can provide daily estimates of the relationship between AOD and $PM_{2.5}$ (similar to mixed effects models) (see the section "Statistical approaches" in this appendix), and can account for the impact of elevated $PM_{2.5}$ layers on this relationship, thereby identifying periods where the AOD is likely dominated by aerosols above the PBL and thus do not provide good data on ground-level $PM_{2.5}$ concentrations. The disadvantages are that it requires the running of a CTM, which can be a complicated and labor-intensive process, and that it is highly dependent on the ability of the CTM to correctly predict aerosol vertical profiles, even when the CTM predictions of AOD are off. However, the use of publicly available CTM output (see the section "Chemical transport model–based approaches" in this appendix) could reduce the labor issues for LMICs.

Errors in AOD can be propagated through the CTM methods by repeating the process for the high and low error bounds on the AOD. However, quantifying the uncertainty in the CTM estimate of the relationship between AOD and surface concentrations is not straightforward, since this relationship is sensitive to errors in multiple aerosol components at multiple vertical levels, as well as errors in relative humidity. Thus, the method used here to account for this error is to assume a constant relative error based on *post hoc* validation of the CTM $PM_{2.5}$ estimates (for example, ±47 percent, based on the "rest of world" results of van Donkelaar, Martin, Brauer, and Boys 2015).

Van Donkelaar, Martin, and Park (2006) used MODIS Dark Target AOD and MISR AOD separately to predict annual mean ground-level $PM_{2.5}$ concentrations over North America using the GEOS-Chem model (v7-02-01, 1° by 1° nested resolution over North America, driven with GEOS-3 meteorology) to predict the relationship between AOD and $PM_{2.5}$. Using MISR AOD explained 34 percent of the variability, and the MISR-predicted values had a mean bias of 3.1 micrograms per cubic meter and a slope of 0.57. Using MODIS Dark Target explained 48 percent of the variability with a mean bias of 5.1 micrograms per cubic meter and a slope of 0.82.

Van Donkelaar and others (2010) used MODIS Dark Target AOD and MISR AOD together to provide long-term average (2001–06) global estimates of ground-level $PM_{2.5}$ concentrations at about 10 kilometers (0.1° by 0.1°) resolution using the GEOS-Chem model (v8-01-04, 2° by 2.5° resolution, driven with GEOS-4 meteorology) to predict the relationship between AOD and $PM_{2.5}$. In the present study, the MODIS and MISR data were filtered to remove observations that had an anticipated bias greater than the larger of ±0.1 or ±20 percent. Remaining MODIS and MISR AOD retrievals were averaged to produce a single value for a grid cell. Over North America, the model explained 59 percent of the observed variability with a slope of 1.07 and a mean bias of –1.75 micrograms per cubic meter, both substantial improvements over the estimates of van Donkelaar, Martin, and Park (2006).

Van Donkelaar and others (2011) used MODIS Dark Target AOD and the GEOS-Chem model (v8-03-01, 2° by 2.5° resolution, driven with GEOS-5 meteorology) to estimate daily $PM_{2.5}$ concentrations during a major biomass burning event around Moscow in the summer of 2010. During this event, the standard MODIS retrieval incorrectly identified some of the aerosol as cloud due to the large AOD values. Relaxing the cloud screening increased MODIS coverage by 21 percent with no evidence of false aerosol detection. GLM $PM_{2.5}$ data were estimated from PM_{10} observations for several nearby sites, because only two sites had $PM_{2.5}$ observations. The satellite product explained 85 percent of the variability in these estimated $PM_{2.5}$ observations with a slope of 1.06.

Geng and others (2015) used the combined MODIS Dark Target and MISR AOD product of van Donkelaar and others (2010) to determine long-term average (2006–12) $PM_{2.5}$ concentrations over China at about 10 km (0.1° by 0.1°) resolution using the GEOS-Chem model (version 9-01-02, 0.5° by 0.667° nested resolution over China, driven with GEOS-5 meteorology) to predict the relationship between AOD and $PM_{2.5}$. Comparison with ground-level $PM_{2.5}$ concentrations observations showed the satellite-based product explained 55 percent of the variability in $PM_{2.5}$ with a slope of 0.77.

Van Donkelaar and others (2015) used satellite AOD observations and the GEOS-Chem model to produce annual average $PM_{2.5}$ estimates (1998–2012) over the globe. The MODIS radiance observations were used in an optimal estimation framework (van Donkelaar and others 2013) to derive $PM_{2.5}$ estimates that were consistent with the GEOS-Chem aerosol scheme for 2004–10, which were then combined with the product of van Donkelaar and others (2010) for 2001–03 to produce a global, decadal average $PM_{2.5}$ estimate at about 10-kilometer resolution. The work of Boys and others (2014), which used AOD from the SeaWIFS and MISR satellites and the GEOS-Chem model to estimate the temporal variation in $PM_{2.5}$, was then applied to this decadal average to produce a 15-year global estimate of ground-level $PM_{2.5}$. Comparisons with decadal mean GLM data over North America gave an *R* of 58 percent with a slope of 0.96 and a 1–σ error of –1 microgram per cubic meter + 15 percent. Over Europe, the comparison gave an *R* of 53 percent with a slope of 0.78 and a 1–σ error of 1 microgram per cubic meter +21 percent, whereas over the rest of the world the *R* was 0.66, the slope was 0.68, and the 1–σ error was –1 microgram per cubic meter + 47 percent.

Sources of AOD-$PM_{2.5}$ relationships from observations

Until recently, few data were available from colocated observations of AOD and ground-level $PM_{2.5}$. The Surface Particulate Matter Network (SPARTAN) was established to address this need (Snider and others 2015). The network includes a global federation of ground-level monitors of hourly $PM_{2.5}$, primarily in highly populated regions in proximity to existing ground-based sun photometers (for example, AERONET sites) that measure AOD. Together, these instruments provide an empirical measure of the $AOD/PM_{2.5}$ ratio that is used to relate satellite AOD retrievals to ground-level $PM_{2.5}$.

The current SPARTAN network is fairly sparse, but adding an AERONET and SPARTAN site to a given city would provide valuable data on the local variation of the $AOD/PM_{2.5}$ ratio at a relatively low cost. However, in the absence of these data, this ratio can be estimated from CTM output.

Sources of aerosol profile data from CTMs

For LMICs to use the CTM approach without having to run CTMs themselves, they need publicly available CTM data sets that can be used to match aerosol profiles with AOD observations. Several potential sources of these data for LMICs are discussed below.

MERRA-2 Reanalysis: The NASA MERRA-2 reanalysis (Provençal and others 2017) provides hourly average AOD at 550 nm both by total and by aerosol component, as well as the surface mass concentration for each component, at a horizontal resolution of 0.5° by 0.625°. However, aerosol vertical profiles are only provided every three hours as instantaneous (snapshot) profiles. MERRA-2 uses the NASA GEOS-5 atmospheric general circulation model (Colarco and others 2014). The aerosol component of MERRA-2 is based on the Goddard Chemistry Aerosol Radiation and Transport Model (GOCART; Chin and others 2000; Colarco and others 2010, 2014; Kim and others 2013). Funded mainly by NASA Earth Science programs, the GOCART model was developed to simulate atmospheric aerosols (including sulfate, black carbon, organic carbon, dust, and sea salt), carbon monoxide, and sulfur gases. The reanalysis has about a two-month latency. All of the data are freely available for download from NASA via MDISC,[48] managed by the NASA Goddard Earth Sciences (GES) Data and Information Services Center (DISC).

NCEP NGAC forecasts: NEMS GFS Aerosol Component (NGAC; Lu, da Silva, and others 2016) is a global inline aerosol forecast system. The forecast model component of the NGAC is the GFS based on the NOAA Environmental Modeling System (NEMS; Black and others 2007, 2009) which, in turn, is based on the common modeling framework using Earth System Modeling Framework (ESMF). The aerosol component of the NGAC is GOCART (Chin and others 2000; Kim and others 2013).

The initial production implementation of NGAC with global-dust-only forecast was implemented in September 2012 (Lu, da Silva, and others 2016). The current NGAC operational forecast produces 120-hour global multispecies forecasts including dust, sea salt, sulfate, and carbonaceous aerosols. The system runs twice daily within the NCEP Production Job Suite at the 0000Z cycle and 1200Z cycle. Although no satellite data assimilation is currently performed, research is underway to add the assimilation of the NOAA VIIRS AOD product to NGAC (Lu, Wei, and others 2016). Output is posted to a 1° by 1° longitude/latitude grid with a 3-hour forecast interval to 120 hours. The files (in grib2 format) include three-dimensional profiles of individual aerosol components (dust, sea salt, sulfate, organic carbon, and black carbon), temperature, and relative humidity, as well as two-dimensional fields including the model-calculated total AOD and the AOD from each aerosol component.

NCAR MOZART-4/GEOS-5 forecasts: NCAR provides output from their MOZART-4/GEOS-5 global air-quality simulations (Emmons and others 2010) for public use, both in NRT and in a historical archive going back to 2007.[49] The representation of tropospheric aerosols in MOZART-4 is based on the work of Tie and others (2001, 2005) and includes the calculation of sulfate, black carbon, primary organic, secondary organic aerosols (SOAs), ammonium nitrate, and sea salt (Lamarque and others 2005). Because only the bulk mass is calculated, a lognormal number distribution is assumed for all aerosols to calculate the surface area (and thus AOD), using a different geometric mean radius and standard deviation for each type of aerosol (based on Chin and others 2002). Sea salt aerosols are included in the model with four size bins (0.1–0.5, 0.5–1.5, 1.5–5, and 5–10 micrometers), and emissions are calculated online (Mahowald, Lamarque, and others 2006). However, the distributions of four sizes of dust (0.05–0.5, 0.5–1.25, 1.25–2.5, and 2.5–5.0 micrometers) are set from monthly mean distributions taken from online calculations in the Community Atmosphere Model (CAM) (Mahowald, Muhs, and others 2006), and thus the model does not include prognostic dust. Hygroscopic growth of the aerosols is determined from the ambient relative humidity, with different rates for each type of aerosol (Chin and others 2002). Washout of all aerosols, except hydrophobic black carbon and organic carbon, is set to 20 percent of the washout rate of nitric acid (HNO_3) (Horowitz 2006; Tie and others 2005). Comparisons of calculated AOD over the ocean to AOD retrievals from the MODIS satellite instrument indicate this is a reasonable washout rate.

The NCAR MOZART-4/GEOS-5 aerosol profiles are provided at a horizontal resolution of 1.875° by 2.5° every six hours. However, the output files include only the bulk mass for each size bin, and so the model AOD has to be calculated from these data by the user.

UK Met Office research ensemble forecast: The UK Met Office provides an operational global forecast at 17-kilometer resolution that includes prognostic and interactive dust for a four-member hybrid ensemble. An additional aerosol forecasting global model, including carbonaceous aerosols and sulfate aerosols, is run as a research product using a copy of the operational model, with aerosol chemistry using the CLASSIC scheme (Bellouin and others 2011; Mulcahy and others 2014). However, the additional aerosols do not feed back to the dynamics. This research product is available only via the principal investigator (Malcolm Brooks, malcolm.e.brooks@metoffice.gov.uk).

ECMWF Copernicus Atmospheric Monitoring Service: The ECMWF Copernicus Atmospheric Monitoring Service (CAMS) produces twice-daily 120-hour global forecasts of total aerosols, aerosol components (dust, sea salt, biomass burning, sulfate), and AOD at about 40-kilometer spatial resolution (0.5625° by 0.5625°) at a three-hour temporal resolution. CAMS also produces the Monitoring Atmospheric Chemistry and Climate (MACC) reanalysis (Toll and others 2015) of AOD and component AOD at about 80-kilometer resolution (1.125° by 1.125°). However, both products currently require purchasing a license from ECMWF and thus may not be the correct solution for many LMICs.

CAMS currently uses a simple bin scheme (IFS-LMD [Integrated Forecast System–Laboratoire de Météorologie Dynamique]) for its near real-time forecasts and the MACC reanalysis. IFS-LMD mainly follows the aerosol treatment in the Laboratoire d'Optique Atmosphérique / LMD-Z Laboratoire de Météorologie Dynamique–Zoom (LOA/LMD-Z) model (Boucher and others 2003; Reddy and others 2005). Five types of tropospheric aerosols are considered: sea salt, dust, organic carbon, black carbon, and sulfate aerosols. Prognostic aerosols of natural origin, such as mineral dust and sea salt, are described using three size bins. Emissions of dust depend on the 10-meter wind, soil moisture, the UV-VIS component of the surface albedo, and the fraction of land covered by vegetation when the surface is snow-free. A correction to the 10-meter wind to account for gustiness is also included (Morcrette and others 2009). Sea-salt emissions are diagnosed using a source function based on work by Guelle and others (2001) and Schulz, de Leeuw, and Balkanski (2004). Sources for the other aerosol types are taken from the Speciated Particulate Emission Wizard (SPEW) and the Emission Database for Global Atmospheric Research (EDGAR) annual or monthly mean climatologies (Dentener and others 2006). Emissions of organic matter, black carbon, and SO_2 linked to fire emissions are obtained using the Global Fire Assimilation System based on MODIS satellite observations of fire radiative power, as described in Kaiser and others (2012).

In addition, the ECMWF forecasts assimilate MODIS AOD data at 550 nanometers in a 4D-Var framework that has been extended to include the aerosol total mixing ratio as an extra control variable (Benedetti and others 2009).

Hybrid approaches

Hybrid approaches are those in which both the output from CTMs and statistical models are used to convert satellite AOD retrievals into predictions of ground-level $PM_{2.5}$. For example, van Donkelaar, Martin, Spurr, and Burnett (2015) used geographically weighted regression (GWR) to relate the errors between the GLM $PM_{2.5}$ observations and their estimate of ground-level $PM_{2.5}$ based on MODIS AOD observations and GEOS-Chem simulations (van Donkelaar and others 2013). They used the Matlab function glmfit, but open-source options in Python are available (for example, the pyglmnet library).

The spatial predictors used were the percentage of urban land cover (at one-kilometer resolution) and the subgrid elevation difference (that is, the difference between the mean elevation for a 0.5° by 0.67° GEOS-Chem grid box, the local elevation from high-resolution maps). In addition, the GEOS-Chem predictions for the relative contributions of nitrate, primary carbonaceous, and secondary organic aerosol to the total $PM_{2.5}$ were used as additional predictors.

GWR was then used to predict the bias between monthly mean ground-level $PM_{2.5}$ estimates and the satellite-derived estimate of ground-level $PM_{2.5}$. This gave spatially varying, but constant with time, bias-correction estimates at a fine resolution (one kilometer) over North America. However, although the bias-correction is constant with time, the uncorrected values vary with time (daily for areas with MODIS coverage) and thus provide daily estimates of ground-level $PM_{2.5}$. The bias-corrected satellite product showed a much higher correlation with surface observation and a better regression slope but had similar RMSE with respect to surface monitors as the uncorrected product.

COMBINING GLM AND SATELLITE DATA

The two most common methods that have been used to simultaneously interpolate satellite-derived $PM_{2.5}$ estimates and GLM data to obtain city-scale $PM_{2.5}$ estimates are co-kriging and simultaneous land-use regression (for example, Alexeeff and others 2015; de Hoogh and others 2016; Lee, Chatfield, and Strawa 2016).

Co-kriging

Like most spatial interpolation schemes, ordinary kriging uses a weighted average of neighboring samples (that is, from a spatially distributed set of ground-level monitors of $PM_{2.5}$) to predict $PM_{2.5}$ concentrations across an urban area. However, in ordinary kriging the weights depend not only on the distance between the prediction location and the measured points, but also on the spatial autocorrelation in the data (Millar and others 2010). Kriging also provides error estimates for its predictions.

Co-kriging is an extension of ordinary kriging that can take advantage of additional data sets or variables, using both the correlations between the monitors and the cross-correlations between the monitor data and the additional data sets to make better predictions. In the application of interest here, the second data set would be the satellite-derived ground-level $PM_{2.5}$ estimates produced using one of the methods outlined earlier in this appendix, but additional data sets, such as meteorological data, can be incorporated via co-kriging (Millar and others 2010; Pearce and others 2009; Wu, Winer, and Delfino 2006). Kriging is generally done using licensed software such as ArcGIS, but open source options (for example, the pyKrige toolkit in Python) also are available.

For LMICs, the advantage of combining satellite data with GLM data via co-kriging is that it potentially reduces the number of GLM stations required to adequately cover an urban area via ordinary kriging. Kriging is most suitable for cases where the GLM network (1) is spatially dense, (2) captures the general spatial variation of $PM_{2.5}$, and (3) adequately covers the edges of the study area. Applying kriging to sparse data will either result in over-smoothed surfaces or over-fitted predictions. Using the more spatially dense satellite estimates of surface level $PM_{2.5}$ in co-kriging could thus address these limitations of kriging for urban areas with sparse GLM networks.

Co-kriging has other potential advantages. It does not require any data beyond the GLM network data and the satellite-derived ground-level $PM_{2.5}$ estimates. In contrast, land-use regression requires substantial additional data (see the sections "Co-kriging" and "Land-use regression" in this appendix). Furthermore, co-kriging can be applied independently at each time point, so that the temporal resolution of the maps produced can be the same as the initial GLM and satellite data, allowing for short-term average (that is, daily) predictions of $PM_{2.5}$ concentrations. In fact, Alexeeff and others (2015) showed that to estimate chronic (annual) exposure to $PM_{2.5}$, kriging applied to annual average data was insufficient to capture the exposure accurately, and thus kriging had to be applied to the daily data and then averaged to provide annual estimates. However, the spatial resolution of the predictions is limited by the spatial density and resolution of the input data sets, making predictions on a one-kilometer scale or finer difficult.

An example of the use of co-kriging to supplement a sparse GLM network was provided by Singh and others (2011), who demonstrated how CTM predictions of ground-level $PM_{2.5}$ concentrations could be used as a secondary variable in a co-kriging interpolation of GLM data to predict PM_{10} concentration over northern Italy. They used 14 PM_{10} monitors and five-kilometer resolution CTM predictions over a 60-kilometer by 60-kilometer domain. Their co-kriging models were able to explain 88–90 percent of the variability in PM_{10} with an RMSE of 16 micrograms per cubic meter, which was lower than the kriging value (20 micrograms per cubic meter). In addition, they found that the predicted maps of PM_{10} concentration were substantially different between kriging and co-kriging, with the co-kriging map capturing more of the expected special variation of PM_{10} in the domain. In addition, Singh, Venkatachalam, and Gautam (2017) used co-kriging to fill in missing MISR AOD data using observations from the MODIS instruments. However, there is no example of co-kriging with satellite-based $PM_{2.5}$ products in the literature.

Land-use regression

Land-use regression (LUR) spatially links GLM network data of $PM_{2.5}$ with other associated variables (such as elevation, distance from roads, population density, and land-use type) to develop finer-scale and more accurate estimates of long-term $PM_{2.5}$ concentrations within an urban area than is possible with kriging (Millar and others 2010). The most predictive variables are identified through least-squares linear regression (Briggs and others 1997). Other studies (for example, Vienneau and others 2013) have incorporated data on meteorology (for example, wind direction; Arain and others 2007; Ryan and others 2008) and satellite observations of aerosols, NO_2, and other pollutants (for example, Lee, Chatfield, and Strawa 2016) into the regression as well. The development of predictor variables from geographic information system (GIS) files is generally performed using licensed software such as ArcGIS, but the regression can be performed using open-source software such as R.

LUR is an attractive option for estimating $PM_{2.5}$ concentrations at a neighborhood scale within urban areas with sparse measurement networks, since the regression can be used to predict concentration at the scale of the land-use data in unmonitored areas. However, the method assesses only spatial variation and thus can predict only long-term average concentrations (for example, monthly or annual). In addition, the method generally requires a significant amount of GIS skill to apply and requires much more input data than co-kriging (see the section "Co-kriging" in this appendix).

Several studies have used satellite estimates of ground-level $PM_{2.5}$ within LUR models. For example, Vienneau and others (2013) used MODIS-derived ground-level $PM_{2.5}$ estimates (at 10-kilometer resolution) as a predictor in a land-use regression model to estimate annual average PM_{10} concentrations over Europe at a resolution of 100 meters. The LUR models using the satellite data had consistently higher correlations and generally lower mean bias and mean absolute errors (MAEs) than the LUR models that did not use satellite data. The models using satellite data to predict annual average PM_{10} generally explained between 35 and 50 percent of the variability in PM_{10} (mean absolute error of 4.6 to 6.0 micrograms per cubic meter).

Lee, Chatfield, and Strawa (2016) adopted a different approach in which LUR is included in the linear mixed-effects statistical model used to relate MODIS Deep Blue AOD to ground-level $PM_{2.5}$ estimates for California (see, for example, the section "Statistical approaches" in this appendix). Separate LUR models were built for each season and trained using three years of data from 142 GLM sites. The land-use predictors included distance from major highways, population density, estimated pollutant emissions from the US EPA National Emission Inventory, elevation, percentage of developed land, distance from the coast, and the air basin in which each measurement site was located. In addition, daily values for AOD, wind speed, relative humidity, temperature, and PBLH from the weather stations nearest to the site were also used as predictors. This addition of temporally varying data with land-use data allows for predictions of daily and annual average $PM_{2.5}$ at the fine spatial scales possible with land-use regression. Lee, Chatfield, and Strawa (2016) found that the model was able to explain 66 percent of the variability of $PM_{2.5}$ over California (MAE of 3.7 micrograms per cubic meter, RMSE of 5.7 micrograms per cubic meter) and that the modeled relationship between AOD and $PM_{2.5}$ varied significantly by day.

De Hoogh and others (2016) used the 10-kilometer resolution satellite-derived estimated of ground-level $PM_{2.5}$ from van Donkelaar, Martin, Spurr, and Burnett (2015) as a predictor in a land-use regression model of $PM_{2.5}$ over western Europe, predicting annual average $PM_{2.5}$ concentrations at a resolution of 100 meters. The model also included satellite estimates of tropospheric NO_2 columns at a 10-kilometer resolution, as well as independent CTM predictions of ground-level $PM_{2.5}$ and NO_2 from the MACC-II ENSEMBLE model, which was the median value of seven individual regional CTMs (CHIMERE, EMEP, EURAD, LOTOS-EUROS, MATCH, MOCAGE, and SILAM). Additional predictors included high-resolution data on land cover, roads, and altitude. They found that the $PM_{2.5}$ LUR models that included both the satellite and CTM estimates of ground-level $PM_{2.5}$ explained about 60 percent of the spatial variation in measured (annual average) $PM_{2.5}$ concentrations, substantially more than was explained by the LUR model without satellite or CTM data (33–38 percent).

Data sets for land-use regression

If an LMIC has detailed GIS data on land use, population density, roads, pollutant emissions, and other fields needed for LUR, those should be used, because they would reflect the best local knowledge regarding the conditions likely to lead to high levels of $PM_{2.5}$. However, since some LMICs may not have these data freely available, high-resolution, global data sets for the predictors used in the LUR studies discussed above were identified in this review:

- Land-cover type can be determined from the 0.5-kilometer MODIS-based Global Land Cover Climatology provided by USGS.[50] This map is based on based on 10 years (2001–10) of Collection 5.1 MCD12Q1 land-cover–type data and used 17 land-cover categories. The map is generated by choosing, for each pixel, the land cover classification with the highest overall confidence from 2001 to 2010 (Broxton and others 2014). These data can also be used to calculate parameters such as the percentage of developed land within a given radius of the site.
- Distance from coasts can be determined using the World Water Bodies GIS layer.[51]
- GIS data on population density[52] and major roads[53] are available through the NASA Socioeconomic Data and Applications Center (SEDAC) hosted at Columbia University.
- The SEDAC Gridded Population of the World, version 4 (GPWv4) Population Density product (CIESIN 2016) consists of estimates of human population density, based on counts consistent with national censuses and population registers, for the years 2000, 2005, 2010, 2015, and 2020. A proportional allocation gridding algorithm, utilizing approximately 12.5 million national and subnational administrative units, is used to assign population values to 30-arc-second (about one-kilometer) grid cells.

- The SEDAC Global Roads Open Access Data Set (CIESIN and ITOS 2013; gROADS) combines the best available roads data by country into a global roads data set, using the UN Spatial Data Infrastructure Transport (UNSDI-T) version 2 as a common data model. Because the data are compiled from multiple sources, the date range for road network representations ranges from the 1980s to 2010 depending on the country (most countries have no confirmed date), and spatial accuracy varies.
- In addition, OpenStreetMap[54] is an open source, community project that maintains data about roads, trails, railway stations, and other landmarks, all over the world. OpenStreetMap emphasizes local knowledge, and contributors use aerial imagery, GPS devices, and low-tech field maps to verify that OpenStreetMap is accurate and up to date. The inspection of these data suggests that the data are more detailed and occasionally more accurate than the gROADS database, but the data sources are also less well documented.
- Topography data can be obtained from the Global Multi-resolution Terrain Elevation Data 2010 (GMTED2010)[55] created by the USGS and the National Geospatial-Intelligence Agency (NGA). This data set has multiple spatial resolutions, 30, 15, and 7.5 arcseconds, and is a replacement for the previous GTOPO30 data.
- Spatially resolved global air pollutant emissions at a 0.1° by 0.1° resolution (about 10 kilometers) are available for the years 2008 and 2010 from the EDGAR-HTAP_V2 emission inventory for several pollutants (CH_4, CO, SO_2, NO_x, nonmethane volatile organic compounds, NH_3, PM_{10}, $PM_{2.5}$, black carbon, and organic carbon).

RECOMMENDATIONS FOR LMICS

These recommendations for the use of satellite observations to supplement GLM data in LMICs (using the typology proposed in chapter 1 and reproduced at the beginning of this appendix) are based on the literature review above:

- For Type I countries, since no GLM data are available, the only possible approach to convert AOD to ground-level $PM_{2.5}$ for these countries is a CTM-based approach. Bias correction of the CTM-based estimates is also not possible for these countries, because there are no data with which to characterize the model bias. Thus, the raw CTM-based estimates of ground-level $PM_{2.5}$ at the native resolution of the satellite AOD product should be used, but the derived ground-level $PM_{2.5}$ values should be assigned a high uncertainty that reflects not only the uncertainty in the AOD but also the estimated uncertainty in the CTM-derived AOD to $PM_{2.5}$ relationship (for example, about 50 percent based on van Donkelaar, Martin, Brauer, and Boys 2015). It is likely in this case that only city-average estimates will be possible, and that the annual estimates will be more reliable than the daily estimates.
- For Type II countries, the small amount of available GLM data (with variable quality) will provide at least some ability to derive a bias estimate for the raw CTM-based estimates of ground-level $PM_{2.5}$ discussed above. However, without improvement in the QA procedures in these countries, it will not be possible to be certain that the bias-corrected estimate is truly more accurate than the raw estimate, and thus both values should be reported and stored. This will allow for reprocessing of the satellite estimates when more rigorous QA

procedures are developed. Similar to Type I countries, it is likely that only city-average estimates will be possible, and that the annual estimates will be more reliable than the daily estimates.

- The GLM data in Type III countries allow both statistical and bias-corrected CTM-based approaches for converting AOD to $PM_{2.5}$ to be considered. CTM-based approaches may still be the best approach for these countries, but statistical approaches should also be tested before making a final decision. If GLM data from several sites exist, but only for a short period, those data can be used to develop seasonally and geographically varying statistical models and/or seasonally and geographically varying bias corrections for CTM-based approaches. If only a single site exists in the urban area, a seasonally varying statistical model and/or bias correction is still possible, but both the model and the bias correction would have to be applied uniformly across the urban area. Land-use regression (LUR; see the section "Land-use regression" in this appendix) can be combined with either technique to provide a finer-scale estimate of annual average ground-level $PM_{2.5}$ (at about 0.5-kilometer resolution) using the data sets described in the section "Data sets for land-use regression" in this appendix, and co-kriging (see the section "Co-kriging" in this appendix) can be used to derive daily estimates across the city.

- Type IV countries will be able to take advantage of the same approaches outlined above for Type III countries. However, the more extensive measurement networks will allow for more accurate estimates of the geographical and seasonal variation in the AOD to $PM_{2.5}$ relationship. Thus, at this point, purely statistical approaches may begin to outperform CTM-based estimates, especially for regions where the model's emission inventories are out of date or otherwise inaccurate.

- In Type V countries, the main use of satellite observations is to help fill in the gaps of the existing monitoring network to cover more of the country's population. These countries also tend to have access to higher-resolution, regional CTM simulations that are customized for their region. Thus, both statistical and CTM-based approaches can be used.

In addition, the following recommendations on data sources and other methodological concerns are made for all countries interested in using satellite data to supplement GLM data:

- All countries should use the satellite product that offers the best balance of accurate AOD and spatial resolution for their country. For example, regions near bright urban or desert surfaces should prefer the MODIS Deep Blue product (see the section "Polar-orbiting satellites" in this appendix), since this has been demonstrated to give better performance in these regions. The extension of this algorithm to VIIRS will help maintain this capability over the next two decades. However, the MAIAC algorithm or the application of the Deep Blue algorithm to geostationary observations may provide better AOD observations in the near future.

- The highest-possible spatial resolution for satellite AOD will likely be between 2 and 4 kilometers (for geostationary observations) or 1 and 10 kilometers (for polar observations). Consequently, finer predictions will be possible only for cases where GLM data exist (via LUR; see the section "Land-use regression" in this appendix).

- Some cities may have poor satellite coverage due to clouds or snow-covered surfaces regardless of the choice of the AOD product. This will reduce the number of days for which a daily estimate is possible and will need to be accounted for in the annual averages.
- If local data on PBLH are available in a city (for example, Ulaanbaatar, Mongolia), the measured PBLH should be used in statistical approaches.
- For countries without the capacity to perform their own meteorological or chemical transport modeling, the freely available global model data sets provided by organizations in the United States for estimates of PBL height (see the section "Statistical approaches" in this appendix), aerosol vertical profiles (see the section "Chemical transport model–based approaches" in this appendix), and other parameters should be used. For long-term studies, the NASA MERRA-2 reanalysis should be used, because this will represent the best estimate of the historical atmospheric state. However, the reanalysis takes two months to produce, and thus for short-term forecasts and advisories the NCEP GFS output (including the NGAC aerosol forecasts) should be used.
- In all cases, the errors from the AOD retrieval, the CTM estimate of the aerosol profiles, and the statistical models used (either directly or for bias correction) should be quantified and used to quantify the error in the derived $PM_{2.5}$ estimate.
- In addition, cases where the NCEP NGAC forecasts and/or the MERRA-2 reanalyses suggest that the AOD is dominated (more than 70 percent) by aerosols above the PBL height should be flagged as potentially poor-quality satellite data, and thus poor-quality ground-level $PM_{2.5}$ estimates.

NOTES

1. https://www.wmo-sat.info/oscar/gapanalyses?variable=6.
2. https://darktarget.gsfc.nasa.gov/.
3. https://deepblue.gsfc.nasa.gov/.
4. This is a formula for the total error, which depends on the AERONET AOD value. At low AOD values, the error will be at least 0.05 whereas at high AOD values the error will be at 15 percent.
5. https://ladsweb.modaps.eosdis.nasa.gov/tools-and-services/#stage_MODIS.pl.
6. https://ladsweb.modaps.eosdis.nasa.gov/search/order/2/MOD04_L2--6,MYD04_L2--6,MOD04_3K--6,MYD04_3K--6.
7. https://worldview.earthdata.nasa.gov.
8. https://ladsweb.modaps.eosdis.nasa.gov/missions-and-measurements/modis/MAIAC_ATBD_v1.pdf.
9. https://ladsweb.modaps.eosdis.nasa.gov/missions-and-measurements/science-domain/maiac/.
10. https://ladsweb.modaps.eosdis.nasa.gov/missions-and-measurements/products/MCD19A2/.
11. https://ladsweb.modaps.eosdis.nasa.gov/archive/allData/6/MCD19A2/.
12. https://ladsweb.modaps.eosdis.nasa.gov/archive/allData/6/MCD19A2/.
13. https://landweb.modaps.eosdis.nasa.gov/cgi-bin/QA_WWW/newPage.cgi.
14. https://landweb.modaps.eosdis.nasa.gov/cgi-bin/QA_WWW/newPage.cgi?fileName=sciTestMenu_C6.
15. https://landweb.modaps.eosdis.nasa.gov/QA_WWW/forPage/C6_test_MOD19.html.
16. https://modis.gsfc.nasa.gov/sci_team/meetings/201606/presentations/land/devadiga.pdf.
17. https://www.star.nesdis.noaa.gov/smcd/emb/viirs_aerosol/index.php/.
18. https://www.avl.class.noaa.gov/saa/products/search?datatype_family=VIIRS_EDR.
19. https://www.avl.class.noaa.gov/saa/products/search?datatype_family=VIIRS_EDR.

20. https://deepblue.gsfc.nasa.gov/data.
21. https://misr.jpl.nasa.gov/getData/accessData/.
22. https://www.ncdc.noaa.gov/cdr/atmospheric/avhrr-aerosol-optical-thickness.
23. https://portal.nccs.nasa.gov/datashare/AVHRRDeepBlue/.
24. https://www.eumetsat.int/what-we-monitor/atmosphere.
25. https://www.eumetsat.int/what-we-monitor/atmosphere.
26. https://www.star.nesdis.noaa.gov/smcd/emb/aerosols/products_geo.php.
27. http://www.ssd.noaa.gov/PS/FIRE/GASP/gasp.html.
28. http://www.icare.univ-lille1.fr/projects/seviri-aerosols.
29. https://www.goes-r.gov.
30. https://www.goes-r.gov/products/baseline-aerosol-opt-depth.html.
31. https://www.goes-r.gov/downloads/users/conferencesAndEvents/2014/GOES-R_Series _Program/04-Laszlo_pres.pdf.
32. https://www.avl.class.noaa.gov/saa/products/welcome.
33. https://www.ncdc.noaa.gov/goes-r-series-satellites/goes-r-series-frequently -asked-questions.
34. https://www.star.nesdis.noaa.gov/jpss/documents/ATBD/ATBD_EPS_Aerosol _AOD_v3.0.1.pdf.
35. https://www.star.nesdis.noaa.gov/star/documents/meetings/2015JPSSAnnual /dayFive/06_Session9_Wolf_STAREnterpriseAlgorithmsPlanJPSSAnnualMeeting 082815.pdf.
36. https://www.jma.go.jp/jma/jma-eng/satellite/index.html.
37. http://www.data.jma.go.jp/mscweb/technotes/msctechrep61-6.pdf.
38. http://www.jmbsc.or.jp/en/meteo-data.html.
39. http://www.jmbsc.or.jp/en/Data/Himawari-8-JMBSC-HP(2017.02.20).pdf.
40. https://www.gaw-wdca.org/.
41. https://www.earlinet.org/index.php?id=earlinet_homepage.
42. http://www-calipso.larc.nasa.gov.
43. http://www.nco.ncep.noaa.gov/pmb/products/gfs.
44. https://weather.gc.ca.
45. https://weather.gc.ca/grib/grib2_glb_25km_e.html.
46. https://www.metoffice.gov.uk/services/data.
47. http://www.noaa.gov/media-release/noaa-to-develop-new-global-weather-model.
48. https://disc.sci.gsfc.nasa.gov/datasets?page=1&keywords=MERRA-2.
49. https://www.acom.ucar.edu/wrf-chem/mozart.shtml.
50. https://lpdaac.usgs.gov/products/mcd12q1v006/.
51. https://www.arcgis.com/home/item.html?id=e750071279bf450cbd510454a80f2e63.
52. http://sedac.ciesin.columbia.edu/data/set/gpw-v4-population-density/data-download.
53. http://sedac.ciesin.columbia.edu/data/set/groads-global-roads-open-access-v1 /data-download.
54. http://www.openstreetmap.org.
55. https://www.usgs.gov/core-science-systems/eros/coastal-changes-and-impacts /gmted2010?qt-science_support_page_related_con=0#qt-science_support_page_related _con.

REFERENCES

Alexeeff, S. E., J. Schwartz, I. Kloog, A. Chudnovsky, P. Koutrakis, and B. A. Coull. 2015. "Consequences of Kriging and Land Use Regression for $PM_{2.5}$ Predictions in Epidemiologic Analyses: Insights into Spatial Variability Using High-Resolution Satellite Data." *Journal of Exposure Science & Environmental Epidemiology* 25 (2): 138–44.

Ao, C. O., D. E. Waliser, S. K. Chan, J. L. Li, B. Tian, F. Xie, and A. J. Mannucci. 2012. "Planetary Boundary Layer Heights from GPS Radio Occultation Refractivity and Humidity Profiles." *Journal of Geophysical Research* 117: D16117.

Arain, M. A., R. Blair, N. Finkelstein, J. R. Brook, T. Sahsuvaroglu, B. Beckerman, L. Zhang, and M. Jerrett. 2007. "The Use of Wind Fields in a Land Use Regression Model to Predict Air Pollution Concentrations for Health Exposure Studies." *Atmospheric Environment* 41 (16): 3453–64.

Atlaskin, E., and T. Vihma. 2012. "Evaluation of NWP Results for Wintertime Nocturnal Boundary-Layer Temperatures over Europe and Finland." *Quarterly Journal of the Royal Meteorological Society* 138: 1440–51.

Bellouin, N., J. Rae, J. A. Jones, C. Johnson, J. Haywood, and O. Boucher. 2011. "Aerosol Forcing in the Climate Model Intercomparison Project (CMIP5) Simulations by HadGEM2-ES and the Role of Ammonium Nitrate." *Journal of Geophysical Research* 116: D20206.

Benedetti, A., J. J. Morcrette, O. Boucher, A. Dethof, R. Engelen, M. Fisher, H. Flentje, N. Huneeus, L. Jones, J. Kaiser, S. Kinne, A. Mangold, M. Razinger, A. J. Simmons, and M. Suttie. 2009. "Aerosol Analysis and Forecast in the European Centre for Medium-Range Weather Forecasts Integrated Forecast System: 2. Data Assimilation." *Journal of Geophysical Research: Atmospheres (1984–2012)* 114: D13205.

Bernard, E., C. Moulin, D. Ramon, D. Jolivet, J. Riedi, and J.-M. Nicolas. 2011. "Description and Validation of an AOT Product over Land at the 0.6 µm Channel of the SEVIRI Sensor Onboard MSG." *Atmospheric Measurement Techniques* 4: 2543–65.

Black, T., H. M. H. Juang, and M. Iredell. 2009. "The NOAA Environmental Modeling System at NCEP." Preprint for the American Meteorological Society, Joint Session 2A.6, "23rd Conference on Weather Analysis and Forecasting/19th Conference on Numerical Weather Prediction," Omaha, NE, June 1–5, 2009.

Black, T., H. M. H. Juang, W. Y. Yang, and M. Iredell. 2007. "An ESMF Framework for NCEP Operational Models." Paper presented at the American Meteorological Society, Joint Session J3.1, "22nd Conference on Weather Analysis and Forecasting/18th Conference on Numerical Weather Prediction," Park City, UT, June 25–29, 2007.

Bosilovich, M. G. 2008. "NASA's Modern Era Retrospective-Analysis for Research and Applications: Integrating Earth Observations." *Earthzine* (September 26, 2008). https://earthzine.org/nasas-modern-era-retrospective-analysis/.

Bosilovich, M. G., S. Akella, L. Coy, R. Cullather, C. Draper, R. Gelaro, R. Kovach, Q. Liu, A. Molod, P. Norris, K. Wargan, W. Chao, R. Reichle, L. Takacs, Y. Vikhliaev, S. Bloom, A. Collow, S. Firth, G. Labow, G. Partyka, S. Pawson, O. Reale, S. D. Schubert, and M. Suarez. 2015. "MERRA-2: Initial Evaluation of the Climate." Technical Report Series on Global Modeling and Data Assimilation no. 43. Greenbelt, MD: NASA Goddard Space Flight Center.

Boucher, O., C. Moulin, S. Belviso, O. Aumont, L. Bopp, E. Cosme, R. von Kuhlmann, M. G. Lawrence, M. Pham, M. S. Reddy, J. Sciare, and C. Venkataraman. 2003. "DMS Atmospheric Concentrations and Sulphate Aerosol Indirect Radiative Forcing: A Sensitivity Study to the DMS Source Representation and Oxidation." *Atmospheric Chemistry and Physics* 3: 49–65.

Boys, B. L., R. V. Martin, A. van Donkelaar, R. J. MacDonell, N. C. Hsu, M. J. Cooper, R. M. Yantosca, Z. Lu, D. G. Streets, Q. Zhang, and S. W. Wang. 2014. "Fifteen-Year Global Time Series of Satellite-Derived Fine Particulate Matter." *Environmental Science & Technology* 48 (19): 11109–18.

Briggs, D. J., S. Collins, P. Elliott, P. Fischer, S. Kingham, E. Lebret, K. Pryl, H. van Reeuwijk, K. Smallbone, and A. van Der Veen. 1997. "Mapping Urban Air Pollution Using GIS: A Regression-Based Approach." *International Journal of Geographical Information Science* 11 (7): 699–718.

Broxton, P. D., X. Zeng, D. Sulla-Menashe, P. A. Troch. 2014. "A Global Land Cover Climatology Using MODIS Data." *Journal of Applied Meteorology and Climatology* 53: 1593–605.

Chatfield, R., M. Sorek-Hamer, A. Lyapustin, and Y. Liu. 2017. "Daily Kilometer-Scale MODIS Satellite Maps of $PM_{2.5}$ Describe Wintertime Episodes." Paper 263193 presented at the "Air and Waste Management Association 2017 Annual Conference and Exhibition," Pittsburgh, PA, June 5–9.

Chin, M., P. Ginoux, S. Kinne, O. Tores, B. N. Holben, B. N. Duncan, R. V. Martin, J. A. Logan, A. Higurashi, and T. Nakajima. 2002. "Tropospheric Aerosol Optical Thickness from the GOCART Model and Comparisons with Satellite and Sun Photometer Measurements." *Journal of Atmospheric Science* 59: 461–83.

Chin, M., R. B. Rood, S. J. Lin, J. F. Müller, and A. M. Thompson. 2000. "Atmospheric Sulfur Cycle Simulated in the Global Model GOCART: Model Description and Global Properties." *Journal of Geophysical Research: Atmospheres* 105 (D20): 24671–87.

CIESIN (Center for International Earth Science Information Network, Columbia University). 2016. "Gridded Population of the World, Version 4 (GPWv4): Population Density." Palisades, NY: NASA Socioeconomic Data and Applications Center (SEDAC). https://doi.org//10.7927/H4NP22DQ.

CIESIN (Center for International Earth Science Information Network, Columbia University) and ITOS (Information Technology Outreach Services, University of Georgia). 2013. "Global Roads Open Access Data Set, Version 1 (gROADSv1)." Palisades, NY: NASA Socioeconomic Data and Applications Center (SEDAC). https://doi.org/10.7927/H4VD6WCT.

Colarco, P., A. da Silva, M. Chin, and T. Diehl. 2010. "Online Simulations of Global Aerosol Distributions in the NASA GEOS-4 Model and Comparisons to Satellite and Ground-Based Aerosol Optical Depth." *Journal of Geophysical Research: Atmospheres* 115: D14207.

Colarco, P. R., E. P. Nowottnick, C. A. Randles, B. Yi, P. Yang, K. M. Kim, J. A. Smith, and C. G. Bardeen. 2014. "Impact of Radiatively Interactive Dust Aerosols in the NASA GEOS-5 Climate Model: Sensitivity to Dust Particle Shape and Refractive Index." *Journal of Geophysical Research: Atmospheres* 119 (2): 753–86.

Cucurull, L., and J. C. Derber. 2008. "Operational Implementation of COSMIC Observations into NCEP's Global Data Assimilation System." *Weather and Forecasting* 23: 702–11.

de Hoogh, K., J. Gulliver, A. van Donkelaar, R. V. Martin, J. D. Marshall, M. J. Bechle, G. Cesaroni, M. C. Pradas, A. Dedele, M. Eeftens, and B. Forsberg. 2016. "Development of West-European $PM_{2.5}$ and NO_2 Land Use Regression Models Incorporating Satellite-Derived and Chemical Transport Modelling Data." *Environmental Research* 151: 1–10.

Dentener, F., S. Kinne, T. Bond, O. Boucher, J. Cofala, S. Generoso, P. Ginoux, S. Gong, J. J. Hoelzemann, and A. Ito, and L. Marelli. 2006. "Emissions of Primary Aerosol and Precursor Gases in the Years 2000 and 1750 Prescribed Data-Sets for AeroCom." *Atmospheric Chemistry and Physics* 6 (12): 4321–44.

Diner, D. J., J. C. Beckert, T. H. Reilly, C. J. Bruegge, J. E. Conel, R. A. Kahn, J. V. Martonchik, T. P. Ackerman, R. Davies, S. A. Gerstl, and H. R. Gordon. 1998. "Multi-Angle Imaging SpectroRadiometer (MISR) Instrument Description and Experiment Overview." *IEEE Transactions on Geoscience and Remote Sensing* 36 (4): 1072–87.

Drüe, C., W. Frey, A. Hoff, and T. Hauf. 2008. "Aircraft-Type Specific Errors in AMDAR Weather Reports from Commercial Aircraft." *Quarterly Journal of the Royal Meteorological Society* 134: 229–39.

Emili, E., A. Lyapustin, Y. Wang, C. Popp, S. Korkin, M. Zebisch, S. Wunderle, and M. Petitta. 2011. "High Spatial Resolution Aerosol Retrieval with MAIAC: Application to Mountain Regions." *Journal of Geophysical Research* 116: D23211.

Emmons, L. K., S. Walters, P. G. Hess, J.-F. Lamarque, G. G. Pfister, D. Fillmore, C. Granier, A. Guenther, D. Kinnison, T. Laepple, J. Orlando, X. Tie, G. Tyndall, C. Wiedinmyer, S. L. Baughcum, and S. Kloster. 2010. "Description and Evaluation of the Model for Ozone and Related Chemical Tracers, Version 4 (MOZART-4)." *Geoscientific Model Development* 3: 43–67.

Fleming, R. J. 1996. "The Use of Commercial Aircraft as Platforms for Environmental Measurements." *Bulletin of the American Meteorological Society* 77: 2229–42.

Gelaro, R., W. McCarty, M. J. Suárez, R. Todling, A. Molod, L. Takacs, C. A. Randles, A. Darmenov, M. G. Bosilovich, R. Reichle, K. Wargan, L. Coy, R. Cullather, C. Draper, S. Akella, V. Buchard, A. Conaty, A. M. da Silva, W. Gu, G.-K. Kim, R. Koster, R. Lucchesi, D. Merkova, J. E. Nielsen, G. Partyka, S. Pawson, W. Putman, M. Rienecker, S. D. Schubert, M. Sienkiewicz, and B. Zhao. 2017. "The Modern-Era Retrospective Analysis for Research and Applications, Version 2 (MERRA-2)." *Journal of Climate* 30 (14): 5419–54.

Geng, G., Q. Zhang, R. V. Martin, A. van Donkelaar, H. Huo, H. Che, J. Lin, and K. He. 2015. "Estimating Long-Term $PM_{2.5}$ Concentrations in China Using Satellite-Based Aerosol Optical Depth and a Chemical Transport Model." *Remote Sensing of Environment* 166: 262–70.

Guelle, W., M. Schulz, Y. Balkanski, and F. Dentener. 2001. "Influence of the Source Formulation on Modeling the Atmospheric Global Distribution of Sea Salt Aerosol." *Journal of Geophysical Research: Atmospheres* 106 (D21): 27509–24.

Guo, P., Y.-H. Kuo, S. V. Sokolovskiy, and D. H. Lenschow. 2011. "Estimating Atmospheric Boundary Layer Depth Using COSMIC Radio Occultation Data." *Journal of the Atmospheric Sciences* 68 (8): 1703–13.

Gupta, P., R. C. Levy, S. Mattoo, L. A. Remer, and L. A. Munchak. 2016. "A Surface Reflectance Scheme for Retrieving Aerosol Optical Depth over Urban Surfaces in MODIS Dark Target Retrieval Algorithm." *Atmospheric Measurement Techniques* 9: 3293–308.

Han, J., and H.-L. Pan. 2011. "Revision of Convection and Vertical Diffusion Schemes in the NCEP Global Forecast System." *Weather and Forecasting* 26: 520–33.

Hegarty, J. D., J. Lewis, L. McGrath-Spangler, J. Henderson, A. J. Scarino, R. Adams-Selin, M. Hicks, R. Ferrare, P. DeCola, and E. Welton. 2018. "Analysis of the Planetary Boundary Layer Height during DISCOVER-AQ Baltimore–Washington, DC with Lidar and High-Resolution WRF Modeling." 2018. *Journal of Applied Meteorology and Climatology* 57: 2679–96.

Hogan, T. F., M. Liu, J. A. Ridout, M. S. Peng, T. R. Whitcomb, B. C. Ruston, C. A. Reynolds, S. D. Eckermann, J. R. Moskaitis, N. L. Baker, J. P. McCormack, K. C. Viner, J. G. McLay, M. K. Flatau, L. Xu, C. Chen, and S. W. Chang. 2014. "The Navy Global Environmental Model." *Oceanography* 27 (3): 116–25.

Holben, B. N., D. Tanre, A. Smirnov, T. F. Eck, I. Slutsker, N. Abuhassan, W. W. Newcomb, J. S. Schafer, B. Chatenet, F. Lavenu, and Y. J. Kaufman. 2001. "An Emerging Ground-Based Aerosol Climatology: Aerosol Optical Depth from AERONET." *Journal of Geophysical Research: Atmospheres* 106 (D11): 12067–97.

Holtz, R. E., S. A. Ackerman, F. W. Nagle, R. Frey, S. Dutcher, R. E. Kuehn, M. A. Vaughan, and B. Baum. 2008. "Moderate Resolution Imaging Spectroradiometer (MODIS) Cloud Detection and Height Evaluation Using CALIOP." *Journal of Geophysical Research* 113: D00A19.

Horowitz, L. W. 2006. "Past, Present, and Future Concentrations of Tropospheric Ozone and Aerosols: Methodology, Ozone Evaluation, and Sensitivity to Aerosol Wet Removal." *Journal of Geophysical Research* 111: D22211.

Hsu, N. C., M. J. Jeong, C. Bettenhausen, A. M. Sayer, R. Hansell, C. S. Seftor, J. Huang, and S. C. Tsay. 2013. "Enhanced Deep Blue Aerosol Retrieval Algorithm: The Second Generation." *Journal of Geophysical Research: Atmospheres* 118 (16): 9296–315.

Hu, X., L. A. Waller, A. Lyapustin, Y. Wang, M. Z. Al-Hamdan, W. L. Crosson, M. G. Estes, S. M. Estes, D. A. Quattrochi, S. J. Puttaswamy, and Y. Liu. 2014. "Estimating Ground-Level $PM_{2.5}$ Concentrations in the Southeastern United States Using MAIAC AOD Retrievals and a Two-Stage Model." *Remote Sensing of Environment* 140: 220–32.

Hu, X., L. A. Waller, A. Lyapustin, Y. Wang, and Y. Liu. 2014. "10-year Spatial and Temporal Trends of $PM_{2.5}$ Concentrations in the Southeastern US Estimated Using High-Resolution Satellite Data." *Atmospheric Chemistry and Physics* 14 (12): 6301–14.

Huang, J., S. Kondragunta, I. Laszlo, H. Liu, L. A. Remer, H. Zhang, S. Superczynski, P. Ciren, B. N. Holben, and M. Petrenko. 2016. "Validation and Expected Error Estimation of Suomi-NPP VIIRS Aerosol Optical Thickness and Ångström Exponent with AERONET." *Journal of Geophysical Research: Atmospheres* 121: 7139–60.

Jackson, J. M., H. Liu, I. Laszlo, S. Kondragunta, L. A. Remer, J. Huang, and H.-C. Huang. 2013. "Suomi-NPP VIIRS Aerosol Algorithms and Data Products." *Journal of Geophysical Research: Atmospheres* 118: 12673–89.

Jordan, N. S., R. M. Hoff, and J. T. Bacmeister. 2010. "Validation of Goddard Earth Observing System-Version 5 MERRA Planetary Boundary Layer Heights Using CALIPSO." *Journal of Geophysical Research: Atmospheres* 115 (D24): D24218.

Kahn, R. A., B. J. Gaitley, J. V. Martonchik, D. J. Diner, K. A. Crean, and B. Holben. 2005. "Multiangle Imaging Spectroradiometer (MISR) Global Aerosol Optical Depth Validation Based on 2 Years of Coincident Aerosol Robotic Network (AERONET) Observations." *Journal of Geophysical Research: Atmospheres* 110: D10S04.

Kahn, R. A., D. L. Nelson, M. J. Garay, R. C. Levy, M. A. Bull, D. J. Diner, J. V. Martonchik, S. R. Paradise, E. G. Hansen, and L. A. Remer. 2009. "MISR Aerosol Product Attributes and Statistical Comparisons with MODIS." *IEEE Transactions on Geoscience and Remote Sensing* 47 (12): 4095–114.

Kaiser, J., A. Heil, M. Andreae, A. Benedetti, N. Chubarova, L. Jones, J.-J. Morcrette, M. Razinger, M. Schultz, M. Suttie, and G. R. van der Werf. 2012. "Biomass Burning Emissions Estimated with a Global Fire Assimilation System Based on Observed Fire Radiative Power." *Biogeosciences* 9 (1): 527–54.

Kim, D., M. Chin, H. Bian, Q. Tan, M. E. Brown, T. Zheng, R. You, T. Diehl, P. Ginoux, and T. Kucsera. 2013. "The Effect of the Dynamic Surface Bareness on Dust Source Function, Emission, and Distribution. *Journal of Geophysical Research: Atmospheres* 118 (2): 871–86.

Knapp, K. R., R. Frouin, S. Kondragunta, and A. Prados. 2005. "Toward Aerosol Optical Depth Retrievals over Land from GOES Visible Radiances: Determining Surface Reflectance." *International Journal of Remote Sensing* 26 (18): 4097–116.

Lamarque, J.-F., J. T. Kiehl, P. G. Hess, W. D. Collins, L. K. Emmons, P. Ginoux, C. Luo, and X. Tie. 2005. "Response of a Coupled Chemistry-Climate Model to Changes in Aerosol Emissions: Global Impact on the Hydrological Cycle and the Tropospheric Burdens of OH, Ozone and NO." *Geophysical Research Letters* 32: L16809.

Lee, H. J., R. B. Chatfield, and A. W. Strawa. 2016. "Enhancing the Applicability of Satellite Remote Sensing for $PM_{2.5}$ Estimation Using MODIS Deep Blue AOD and Land Use Regression in California, United States." *Environmental Science & Technology* 50 (12): 6546–55.

Levy, R. C., S. Mattoo, L. A. Munchak, L. A. Remer, A. M. Sayer, F. Patadia, and N. C. Hsu. 2013. "The Collection 6 MODIS Aerosol Products over Land and Ocean." *Atmospheric Measurement Techniques* 6: 2989–3034.

Levy, R. C., L. A. Munchak, S. Mattoo, F. Patadia, L. A. Remer, and R. E. Holz. 2015. "Towards a Long-Term Global Aerosol Optical Depth Record: Applying a Consistent Aerosol Retrieval Algorithm to MODIS and VIIRS-Observed Reflectance." *Atmospheric Measurement Techniques* 8: 4083–110.

Lewis, J. R., E. J. Welton, A. M. Molod, and E. Joseph 2013. "Improved Boundary Layer Depth Retrievals from MPLNET." *Journal of Geophysical Research: Atmospheres* 118: 9870–79.

Li, Y., Y. Xue, G. de Leeuw, C. Li, L. Yang, T. Hou, and F. Marir. 2013. "Retrieval of Aerosol Optical Depth and Surface Reflectance over Land from NOAA AVHRR Data." *Remote Sensing of Environment* 133: 1–20.

Lu, C.-H., A. da Silva, J. Wang, S. Moorthi, M. Chin, P. Colarco, Y. Tang, P. S. Bhattacharjee, S.-P. Chen, H.-Y. Chuang, H.-M. H. Juang, J. McQueen, and M. Iredell. 2016. "The Implementation of NEMS GFS Aerosol Component (NGAC) Version 1.0 for Global Dust Forecasting at NOAA/NCEP." *Geoscientific Model Development* 9: 1905–19.

Lu, C.-H., S.-W. Wei, S. Kondragunta, Q. Zhao, J. McQueen, J. Wang, and P. Bhattacharjee. 2016. "NCEP Aerosol Data Assimilation Update: Improving NCEP Global Aerosol Forecasts Using JPSS-NPP VIIRS Aerosol Products." Paper presented at the "8th International Cooperative for Aerosol Prediction." http://icap.atmos.und.edu/ICAP8/Day3/Lu_NCEP_ThursdayAM.pdf.

Lv, B., Y. Hu, H. H. Chang, A. G. Russell, and Y. Bai. 2016. "Improving the Accuracy of Daily $PM_{2.5}$ Distributions Derived from the Fusion of Ground-Level Measurements with Aerosol Optical Depth Observations, a Case Study in North China." *Environmental Science & Technology* 50 (9): 4752–59.

Lyapustin, A., Y. Wang, I. Laszlo, R. Kahn, S. Korkin, L. Remer, R. Levy, and J. S. Reid. 2011. "Multiangle Implementation of Atmospheric Correction (MAIAC): 2. Aerosol Algorithm." *Journal of Geophysical Research* 116: D03211.

Mahowald, N., J.-F. Lamarque, X. Tie, and E. Wolff. 2006. "Sea Salt Aerosol Response to Climate Change: Last Glacial Maximum, Pre-Industrial and Doubled Carbon Dioxide Climates." *Journal of Geophysical Research* 111: D05303.

Mahowald, N. M., D. R. Muhs, S. Levis, P. J. Rasch, M. Yoshioka, C. S. Zender, and C. Luo. 2006. "Change in Atmospheric Mineral Aerosols in Response to Climate: Last Glacial Period, Preindustrial, Modern, and Doubled Carbon Dioxide Climates." *Journal of Geophysical Research* 111: D10202.

Martins, V. S., A. Lyapustin, L. A. S. de Carvalho, C. C. F. Barbosa, and E. M. L. M. Novo. 2017. "Validation of High-Resolution MAIAC Aerosol Product over South America." *Journal of Geophysical Research: Atmospheres* 122: 7537–59.

Martins, J. P. A., J. Teixeira, P. M. M. Soares, P. M. A. Miranda, B. H. Kahn, V. T. Dang, F. W. Irion, E. J. Fetzer, and E. Fishbein. 2010. "Infrared Sounding of the Trade-Wind Boundary Layer: AIRS and the RICO Experiment." *Geophysical Research Letters* 37 (24): L24806.

Martonchik, J. V., D. J. Diner, R. Kahn, B. Gaitley, and B. N. Holben. 2004. "Comparison of MISR and AERONET Aerosol Optical Depths over Desert Sites." *Geophysical Research Letters* 31 (16): L16102.

McGrath-Spangler, E. L., and A. S. Denning. 2012. "Estimates of North American Summertime Planetary Boundary Layer Depths Derived from Space-Borne Lidar." *Journal of Geophysical Research: Atmospheres* 117: D15101.

McGrath-Spangler, E. L., and A. S. Denning. 2013. "Global Seasonal Variations of Midday Planetary Boundary Layer Depth from CALIPSO Space-Borne LIDAR." *Journal of Geophysical Research: Atmospheres* 118: 1226–33.

Mei, L., Y. Xue, G. de Leeuw, T. Holzer-Popp, J. Guang, Y. Li, L. Yang, H. Xu, X. Xu, C. Li, Y. Wang, C. Wu, T. Hou, X. He, J. Liu, J. Dong, and Z. Chen. 2012. "Retrieval of Aerosol Optical Depth over Land Based on a Time Series Technique Using MSG/SEVIRI Data." *Atmospheric Chemistry and Physics* 12: 9167–85.

Millar, G., T. Abel, J. Allen, P. Barn, M. Noullett, J. Spagnol, and P. L. Jackson. 2010. "Evaluating Human Exposure to Fine Particulate Matter Part II: Modeling." *Geography Compass* 4 (7): 731–49.

Morcrette, J.-J., O. Boucher, L. Jones, D. Salmond, P. Bechtold, A. Beljaars, A. Benedetti, A. Bonet, J. Kaiser, M. Razinger, M. Schulz, S. Serrar, A. J. Simmons, M. Sofiev, M. Suttie, A. M. Tompkins, and A. Untch. 2009. "Aerosol Analysis and Forecast in the European Centre for Medium-Range Weather Forecasts Integrated Forecast System: Forward Modeling." *Journal of Geophysical Research: Atmospheres (1984–2012)* 114 (D6): D06206.

Mulcahy, J. P., D. N. Walters, N. Bellouin, and S. F. Milton. 2014. "Impacts of Increasing the Aerosol Complexity in the Met Office Global Numerical Weather Prediction Model." *Atmospheric Chemistry and Physics* 14: 4749–78.

Munchak, L. A., R. C. Levy, S. Mattoo, L. A. Remer, B. N. Holben, J. S. Schafer, C. A. Hostetler, and R. A. Ferrare. 2013. "MODIS 3 km Aerosol Product: Applications over Land in an Urban/Suburban Region." *Atmospheric Measurement Techniques* 6: 1747–59.

Oo, M. M., M. Jerg, E. Hernandez, A. Picon, B. M. Gross, F. Moshary, and S. A. Ahmed. 2010. "Improved MODIS Aerosol Retrieval Using Modified VIS/SWIR Surface Albedo Ratio over Urban Scenes." *IEEE Transactions on Geoscience and Remote Sensing* 48 (3): 983–1000.

Pearce, J. L., S. L. Rathbun, M. Aguilar-Villalobos, and L. P. Naeher. 2009. "Characterizing the Spatiotemporal Variability of $PM_{2.5}$ in Cusco, Peru Using Kriging with External Drift. *Atmospheric Environment* 43 (12): 2060–69.

Provençal, S., V. Buchard, A. M. da Silva, R. Leduc, N. Barrette, E. Elhacham, and S. H. Wang. 2017. "Evaluation of $PM_{2.5}$ Surface Concentrations Simulated by Version 1 of NASA's MERRA Aerosol Reanalysis over Israel and Taiwan." *Aerosol and Air Quality Research* 17: 253–61.

Reddy, M. S., O. Boucher, N. Bellouin, M. Schulz, Y. Balkanski, J.-L. Dufresne, and M. Pham. 2005. "Estimates of Global Multicomponent Aerosol Optical Depth and Direct Radiative Perturbation in the Laboratoire de Météorologie Dynamique General Circulation Model." *Journal of Geophysical Research: Atmospheres (1984–2012)* 110 (D10): D10S16.

Remer, L. A., S. Mattoo, R. C. Levy, and L. A. Munchak. 2013. "MODIS 3 km Aerosol Product: Algorithm and Global Perspective." *Atmospheric Measurement Techniques* 6: 1829–44.

Rienecker, M. M., M. J. Suarez, R. Todling, J. Bacmeister, L. Takacs, H.-C. Liu, W. Gu, M. Sienkiewicz, R. D. Koster, R. Gelaro, I. Stajner, and J. E. Nielsen. 2008. "The GEOS-5 Data Assimilation System—Documentation of Versions 5.0.1, 5.1.0, and 5.2.0." *NASA GSFC Technical Report Series on Global Modeling and Data Assimilation* vol. 27, NASA/TM-2007-104606.

Ryan, P. H., G. K. LeMasters, L. Levin, J. Burkle, P. Biswas, S. Hu, S. Grinshpun, and T. Reponen. 2008. "A Land-Use Regression Model for Estimating Microenvironmental Diesel Exposure Given Multiple Addresses from Birth through Childhood." *Science of the Total Environment* 404 (1): 139-47.

Sayer, A. M., N. C. Hsu, J. Lee, N. Carletta, S.-H. Chen, and A. Smirnov. 2017. "Evaluation of NASA Deep Blue/SOAR Aerosol Retrieval Algorithms Applied to AVHRR Measurements." *Journal of Geophysical Research: Atmospheres* 122: 9945–67.

Sayer, A. M., L. A. Munchak, N. C. Hsu, R. C. Levy, C. Bettenhausen, and M.-J. Jeong. 2014. "MODIS Collection 6 Aerosol Products: Comparison between Aqua's e-Deep Blue, Dark Target, and "Merged" Data Sets, and Usage Recommendations." *Journal of Geophysical Research: Atmospheres* 119: 13965–89. https://doi.org/10.1002/2014JD022453.

Schneider, T. 2016. "Next Generation Global Prediction System (NGGPS) Phase 2 Atmospheric Dynamic Core Evaluation." Presentation for UMAC, June 22, 2016. https://www.weather.gov/media/sti/nggps/Phase%202%20Dycore%20Evaluation%20Briefing%2022%20June%202016%20UMAC%20Final%20for%20posting%2006272016.pdf.

Schulz, M., G. de Leeuw, and Y. Balkanski. 2004. "Sea-Salt Aerosol Source Functions and Emissions." In *Emissions of Atmospheric Trace Compounds*, edited by C. Granier, P. Artaxo, and C. E. Reeves, 333–59. Dordrecht, the Netherlands: Kluwer.

Seidel, D. J., C. O. Ao, and K. Li. 2010. "Estimating Climatological Planetary Boundary Layer Heights from Radiosonde Observations: Comparison of Methods and Uncertainty Analysis." *Journal of Geophysical Research* 115 (D16): D16113.

Seidel, D. J., Y. Zhang, A. C. M. Beljaars, J.-C. Golaz, A. R. Jacobson, and B. Medeiros. 2012. "Climatology of the Planetary Boundary Layer over the Continental United States and Europe." *Journal of Geophysical Research* 117: D17106.

Singh, V., C. Carnevale, G. Finzi, E. Pisoni, and M. Volta. 2011. "A Cokriging Based Approach to Reconstruct Air Pollution Maps, Processing Measurement Station Concentrations and Deterministic Model Simulations." *Environmental Modelling & Software* 26 (6): 778–86.

Singh, M. K., P. Venkatachalam, and R. Gautam. 2017. "Geostatistical Methods for Filling Gaps in Level-3 Monthly-Mean Aerosol Optical Depth Data from Multi-Angle Imaging Spectroradiometer." *Aerosol and Air Quality Research* 17: 1963–74.

Snider, G., C. L. Weagle, R. V. Martin, A. van Donkelaar, K. Conrad, D. Cunningham, C. Gordon, M. Zwicker, C. Akoshile, P. Artaxo, and N. X. Anh. 2015. "SPARTAN: A Global Network to Evaluate and Enhance Satellite-Based Estimates of Ground-Level Particulate Matter for Global Health Applications." *Atmospheric Measurement Techniques* 8: 505–21.

Sorek-Hamer, M., I. Kloog, P. Koutrakis, A. W. Strawa, R. Chatfield, A. Cohen, W. L. Ridgway, and D. M. Broday. 2015. "Assessment of PM$_{2.5}$ Concentrations over Bright Surfaces Using MODIS Satellite Observations." *Remote Sensing of Environment* 163: 180–85.

Sorek-Hamer, M., A. W. Strawa, R. B. Chatfield, R. Esswein, A. Cohen, and D. M. Broday. 2013. "Improved Retrieval of PM$_{2.5}$ from Satellite Data Products Using Non-Linear Methods." *Environmental Pollution* 182: 417–23.

Strawa, A. W., R. B. Chatfield, M. Legg, B. Scarnato, and R. Esswein. 2013. "Improving Retrievals of Regional Fine Particulate Matter Concentrations from Moderate Resolution Imaging Spectroradiometer (MODIS) and Ozone Monitoring Instrument (OMI) Multisatellite Observations." *Journal of the Air & Waste Management Association* 63 (12): 1434–46.

Sun, W., G. Videen, S. Kato, B. Lin, C. Lukashin, and Y. Hu. 2011. "A Study of Subvisual Clouds and Their Radiation Effect with a Synergy of CERES, MODIS, CALIPSO, and AIRS Data." *Journal of Geophysical Research: Atmospheres* 116: D22207.

Superczynski, S. D., S. Kondragunta, and A. I. Lyapustin. 2017. "Evaluation of the Multi-Angle Implementation of Atmospheric Correction (MAIAC) Aerosol Algorithm through Intercomparison with VIIRS Aerosol Products and AERONET." *Journal of Geophysical Research: Atmospheres* 122: 3005–22. https://doi.org/10.1002/2016JD025720.

Tie, X., G. Brasseur, L. Emmons, L. Horowitz, and D. Kinnision. 2001. "Effects of Aerosols on Tropospheric Oxidants: A Global Model Study." *Journal of Geophysical Research* 106: 2931–64.

Tie, X., S. Madronich, S. Walters, D. Edwards, P. Ginoux, N. Mahowald, R. Zhang, C. Luo, and G. Brasseur. 2005. "Assessment of the Global Impact of Aerosols on Tropospheric Oxidants." *Journal of Geophysical Research* 110: D03204.

Toll, V., K. Reis, R. Ots, M. Kaasik, A. Männik, M. Prank, and M. Sofiev. 2015. "SILAM and MACC Reanalysis Aerosol Data Used for Simulating the Aerosol Direct Radiative Effect with the NWP Model HARMONIE for Summer 2010 Wildfire Case in Russia." *Atmospheric Environment* 121: 75–85.

van Donkelaar, A., R. V. Martin, M. Brauer, and B. L. Boys. 2015. "Use of Satellite Observations for Long-Term Exposure Assessment of Global Concentrations of Fine Particulate Matter." *Environmental Health Perspectives* 123 (2): 135–43.

van Donkelaar, A., R. V. Martin, M. Brauer, R. Kahn, R. Levy, C. Verduzco, and P. J. Villeneuve. 2010. Global Estimates of Ambient Fine Particulate Matter Concentrations from Satellite-Based Aerosol Optical Depth: Development and Application." *Environmental Health Perspectives* 118 (6): 847–55.

van Donkelaar, A., R. V. Martin, and R. J. Park. 2006. "Estimating Ground-Level $PM_{2.5}$ Using Aerosol Optical Depth Determined from Satellite Remote Sensing." *Journal of Geophysical Research: Atmospheres* 111: D21201.

van Donkelaar, A., R. V. Martin, R. J. Spurr, and R. T. Burnett. 2015. "High-Resolution Satellite-Derived $PM_{2.5}$ from Optimal Estimation and Geographically Weighted Regression over North America." *Environmental Science & Technology* 49 (17): 10482–91.

van Donkelaar, A., R. V. Martin, R. J. Spurr, E. Drury, L. A. Remer, R. C. Levy, and J. Wang. 2013. "Optimal Estimation for Global Ground-Level Fine Particulate Matter Concentrations." *Journal of Geophysical Research: Atmospheres* 118 (11): 5621–36.

Vienneau, D., K. de Hoogh, M. J. Bechle, R. Beelen, A. van Donkelaar, R. V. Martin, D. B. Millet, G. Hoek, and J. D. Marshall. 2013. "Western European Land Use Regression Incorporating Satellite- and Ground-Based Measurements of NO_2 and PM_{10}." *Environmental Science & Technology* 47 (23): 13555–64.

Winker, D. M., W. H. Hunt, and M. J. McGill. 2007. "Initial Performance Assessment of CALIOP." *Geophysical Research Letters* 34 (19): L19803.

Winker, D. M., M. A. Vaughan, A. Omar, Y. Hu, K. A. Powell, Z. Liu, W. H. Hunt, and S. A. Young. 2009. "Overview of the CALIPSO Mission and CALIOP Data Processing Algorithms." *Journal of Atmospheric and Oceanic Technology* 26 (11): 2310–23.

Wu, J., A. M. Winer, and R. J. Delfino. 2006. "Exposure Assessment of Particulate Matter Air Pollution before, during, and after the 2003 Southern California Wildfires." *Atmospheric Environment* 40 (18): 3333–48.

Zhang, H., S. Kondragunta, I. Laszlo, H. Liu, L. A. Remer, J. Huang, S. Superczynski, and P. Ciren. 2016. "An Enhanced VIIRS Aerosol Optical Thickness (AOT) Retrieval Algorithm over Land Using a Global Surface Reflectance Ratio Database." *Journal of Geophysical Research: Atmospheres* 121: 10717–38.

Zhu, Y., J. C. Derber, R. J. Purser, B. A. Ballish, and J. Whiting. 2015. "Variational Correction of Aircraft Temperature Bias in the NCEP's GFI Analysis System." *Monthly Weather Review* 143: 3774–803.

APPENDIX B

Converting Satellite Aerosol Optical Depth to Ground-Level PM$_{2.5}$

In this task of the research underlying this report, data from publicly available ground-level-monitoring (GLM) networks (as included in the OpenAQ database) were used to identify promising new pathways for low- and middle-income countries (LMICs) to use satellite observations in their air-quality monitoring and forecasting. As called for in the Terms of Reference for this report, this work "(i) compare[d] and contrast[ed] GLMs in selected LMICs; and (ii) examine[d] novel methods for improving the conversion of aerosol optical depth (AOD) and other remotely sensed data to concentration estimates of surface-level particulate matter with an aerodynamic diameter less than or equal to 2.5 microns (PM$_{2.5}$) in environments typical of LMIC conditions."

The GLM data sets for testing the satellite approaches were data for three cities currently included in the OpenAQ database, which include near real-time (NRT) PM$_{2.5}$ observations from local government air-quality agencies and US embassy and consulate PM$_{2.5}$ data (table B.1). Kriging of the local government data was used to estimate PM$_{2.5}$ values for comparison with the US embassy and consulate PM$_{2.5}$ values, and discrepancy between these estimates was used as a quality check on the local government data.

The use of various satellite data sets was examined for these three cities. The literature review (appendix A) suggested that the Moderate-Resolution Imaging Spectroradiometer (MODIS) combined Deep Blue–Dark Target product was likely the best in terms of the coverage in urban areas, the public availability of the data set, and the long-term global record for training statistical models. The Aqua and Terra data sets were tested separately because they have different overhead crossing times, and the one that gave the best fit with the GLM network data was determined. Two other satellite data sets were also tested. In Delhi, India, the Visible Infrared Imaging Radiometer Suite (VIIRS) six-kilometer resolution AOD retrievals was tested as a potential alternative to the MODIS data. Both MODIS instruments are well past their design lifetimes, and new VIIRS instruments will be launched over the next decade as part of the US JPSS program, so the VIIRS product is more likely to be available in the future than MODIS. The Spinning Enhanced Visible and Infrared Imager (SEVIRI) data set was also used as an example of the capability of next-generation geostationary satellites. Because SEVIRI did not cover the three initially selected test cities, Accra, Ghana, was added as an additional test city for this satellite.

TABLE B.1 **Three focus cities where OpenAQ aggregates real-time local government particulate matter data overlap with US Diplomatic Post measurements**

CITY	COUNTRY	NUMBER OF STATIONS	ORIGINATING DATA SOURCE
Delhi	India	11	Central Pollution Control Board http://www.cpcb.gov.in/CAAQM/mapPage/frmindiamap.aspx
Lima	Peru	10	Servicio Nacional de Meteorología e Hidrología del Perú http://peruclima.pe/?p=calidad-de-aire
Ulaanbaatar	Mongolia	7	Agaar.mn/National Agency of Meteorology and Environment Monitoring http://agaar.mn/

Source: World Bank.

However, the GLM data for Accra are not yet available in OpenAQ, and thus this work merely compared the SEVIRI results with those using the MODIS products for this city. For each satellite data source, all recommended quality flags were applied to filter the AOD data, accepting only the highest quality data from each source.

One statistical approach and one chemical-transport-model (CTM) –based approach were tested for each city (except Accra and Lima, where only the CTM-based approaches were possible). The statistical approach used the generalized additive model (GAM) approach of Sorek-Hamer and others (2013), modified to use the ratio of the AOD to planetary boundary layer (PBL) as the primary predictor, as in Chatfield and others (2017). The CTM approach used AODs and aerosol vertical profiles from the Modern Era Retrospective-analysis for Research and Applications (MERRA-2) reanalysis to derive a ground-level $PM_{2.5}$ estimate following the methods of van Donkelaar and others (2010). Geographically weighted regression (GWR) was then attempted to relate the errors between the GLM $PM_{2.5}$ observations and the two satellite-based estimates of ground-level $PM_{2.5}$, as in van Donkelaar and others (2015).

METHODOLOGY

Satellite data

MODIS and VIIRS data

MODIS and VIIRS products were collected for a two-year period (2016 and 2017) for Delhi, India; Lima, Peru; and Ulaanbaatar, Mongolia. The MODIS data are available through the National Aeronautics and Space Administration's (NASA's) Distributed Active Archive Center (DAAC),[1] where all the MODIS product files for each year and for the Aqua and Terra platforms are available for download. These files are simple to download and inspect, check the spatial range covered in the granule, and keep the granules that cover the region(s) of interest. This is very time consuming but can be automated, as was done for this work.

For VIIRS, the Comprehensive Large Array-Data Stewardship System (CLASS)[2] web-based system was used for ordering data. Here the user has to work through the CLASS graphical user interface and wait for the order to be available; the process cannot be automated further. The process is the following:

assign a date range and draw a box around the region of interest (note the box configuration can be saved to be reloaded for future use to ensure the same region is being searched for each time range). Note that for VIIRS to specifically request that the Earth location data be included (that is, the latitudes and longitudes are not included with the aerosol product), the user must request a final product file that is a merger of the VIIRS geolocation product (GAERO) and the aerosol product (VAOOO).

The AOD data files for the MODIS product contain several different "products" based on the Dark Target and Deep Blue algorithms. The combined Dark Target–Deep Blue product was chosen for this project. This seems to satisfy the requirements and needs of this project the best by providing the greatest possible coverage over urban areas. For VIIRS there is only a single AOD product.

For both VIIRS and MODIS, a region around the focus sites is extracted and a subset database is created for further study. The region extracted is defined for a 2° by 2° region surrounding the center of each city. Table B.2 lists the variables included in the postprocessed MODIS and VIIRS files. Table B.3 discusses how these variables were mapped from the variables in the MODIS and VIIRS files from the NASA DAAC and CLASS, respectively. Note the different definitions for the MODIS and VIIRS products for some of the data products in the focus site files. An uncertainty estimate was also added to the subset files based on the analysis of Remer and others (2013) following the equation

$$AOT_{UNC} = AOT_{BIAS} + AOT_{RMS} \times AOT, \tag{B.1}$$

where AOT_{BIAS} = 0.03 over ocean and 0.05 over land, and AOT_{RMS} = 0.05 over ocean and 0.15 over land, and BIAS = bias, RMS = root mean square, and UNC = uncertainty.

Figure B.1 illustrates the geographic coverage for an approximately 1° by 1° box around each of the focus regions, which was the area used for the analysis.

TABLE B.2 **Contents of the subsampled aerosol-optical-depth product files**

PRODUCT NAME	DESCRIPTION
Algorithm_Flag	MODIS: Defines the algorithm used for generating the product, 0: Dark Target, 1: Deep Blue, 2: Mixed
	VIIRS: 0: Dark Target
AOD_550	MODIS: AOD at 0.55 μm for combined Dark Target–Deep Blue products
	VIIRS: Dark Target product
Error_Estimate	The expected error for each product
Land_Sea_Flag	[0,1,2]: Land/sea/coast flag, 0: ocean, 1: land, 2: coastal
Latitude	Pixel latitude degree
Longitude	Pixel longitude degree
nPix	Number of pixels in file
ObsTimeInSec	Observation time in seconds from start of day
QA_Flag	MODIS: Combined Dark Target–Deep Blue confidence flag (0: no confidence, 1: marginal, 2: good, 3: very good)
	VIIRS: Dark Target aerosol confidence–quality flag (0: not produced, 1: low, 2: medium, 3: high)
Scan_Start_Time	Start time of scan: the pixel is in seconds since January 1, 1993

Source: World Bank.
Note: Note the different product meaning for VIIRS (Visible Infrared Imaging Radiometer Suite) and MODIS (Moderate-Resolution Imaging Spectroradiometer) products. AOD = aerosol optical depth; QA = quality assurance.

TABLE B.3 Mapping between the parsed data file variable names and the aerosol-optical-depth products in the L2 files

PRODUCT NAME	INSTRUMENT	IN INSTRUMENT FILE
Algorithm_Flag	MODIS	AOD_550_Dark_Target_Deep_Blue_Combined_Algorithm_Flag
	VIIRS	Algorithm_Flag_Land
AOD_550	MODIS	AOD_550_Dark_Target_Deep_Blue_Combined
	VIIRS	AerosolOpticalDepth_at_550 nm
QA_Flag	MODIS	AOD_550_Dark_Target_Deep_Blue_Combined_QA_Flag
	VIIRS	QF1_VIIRSAEROEDR: first 2 bits, AOT Quality
		High: Number of good-quality pixel AOT retrievals more than 16 (1/4 the total number of pixels in aggregated horizontal cell)
		Medium: Number of good-quality retrievals less than or equal to 16 and the number of good/degraded quality retrievals greater than or equal to 16
		Low: Number of good/degraded-quality retrievals less than 16
		Not produced: No good/degraded-quality retrievals, neither land nor seawater dominant (number of land or ocean pixels less than half the number of good/degraded pixels in the horizontal cell), ellipsoid fill in the geolocation, night scan, has a pixel with a solar zenith angle between 80° and 85° but no pixel with a solar zenith angle between 65° and 80°

Source: World Bank.
Note: AOD = aerosol optical depth; AOT = aerosol optical thickness; L2 = level 2; MODIS = Moderate-Resolution Imaging Spectroradiometer; QA = quality assurance; VIIRS = Visible Infrared Imaging Radiometer Suite.

FIGURE B.1

Illustration of the regions covered by a 1° by 1° box around each focus region

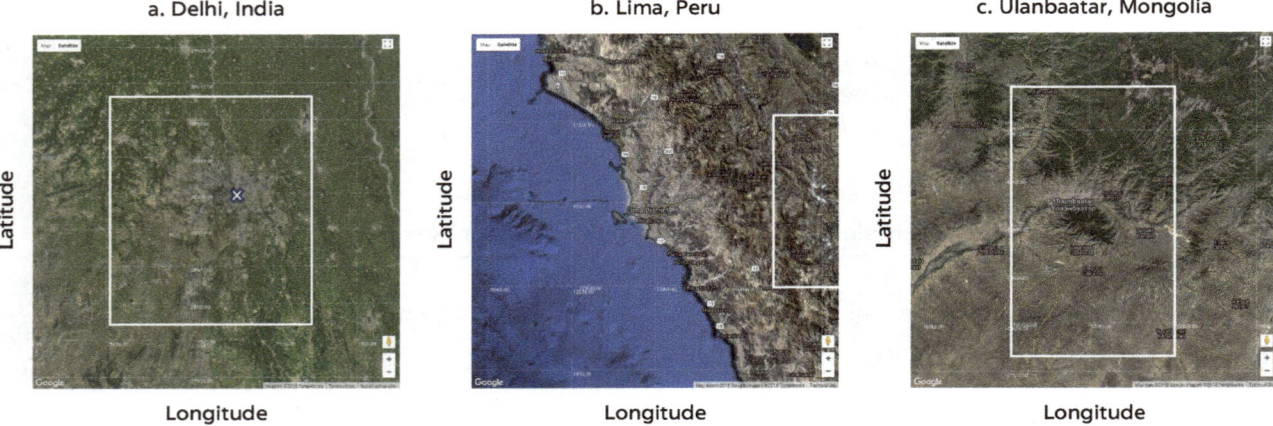

Source: World Bank, produced using Google Maps and Esri ArcGIS.

The image presented for each city covers approximately the circa 2° by 2° box in the subset files.

SEVIRI data

The SEVIRI–Meteosat Second Generation (MSG) Aerosol Over Land (SMAOL) product data were provided through ICARE.[3] Using a Python script to access the ICARE FTP server, the HDF4-formatted, near real-time SMAOL v.1.3.6 product files were collected and data values within ±2° in latitude and longitude of Accra were extracted. The extracted information (72 by 69 pixels at three-kilometer resolution) was saved in netCDF4 files (see table B.4). The latitude and longitude for the SEVIRI grid was obtained from the file "MSG+0000.3km.hdf." The SMAOL product files "SMAOL-AOT-NRT_V1.3.6_*. hdf" include the AOD retrievals for each of six aerosol models (1: Continental WMO, 2: Moderately Absorbing, 3: Urban Industrial, 4: Smoke, 5: Spheroidal Dust, and 6: Maritime) in addition to the best model AOD and

TABLE B.4 **Region of interest data fields in the SEVIRI netCDF output files**

STANDARD FIELDS	DESCRIPTION/SOURCE
AOT_DQF	AOT quality flag: 0 = Clean AOT, 1 = Best AOT, 2 = No AOT
	(derived from AOT550_clean from SMAOL-AOT-NRT_V1.3.6_*.hdf)
AOT_LND	AOT over land
	from AOT550_Best_Model_unfiltered from
	SMAOL-AOT-NRT_V1.3.6_*.hdf
AOT_UNC	AOT uncertainty estimated based on published performance
CloudMask	Four-Level Cloud Mask from SMAOL quality flags
	(extracted from Quality_Flags from SMAOL-AOT-NRT_V1.3.6_*.hdf)
Landmask	Landmask: 0 = Land, 1 = Ocean, from SMAOL quality flags
	(extracted from Quality_Flags from SMAOL-AOT-NRT_V1.3.6_*.hdf)
Latitude	Latitude from MSG+0000.3km.hdf
Longitude	Longitude from MSG+0000.3km.hdf
SMAOL DIAGNOSTIC FIELDS	**DESCRIPTION/SOURCE**
_AOT_LND_Aerosol_Model	Aerosol_Model from SMAOL-AOT-NRT_V1.3.6_*.hdf
_AOT_LND_Model1	AOT550_mode_1_unfiltered from
	SMAOL-AOT-NRT_V1.3.6_*.hdf
_AOT_LND_Model2	AOT550_mode_2_unfiltered from
	SMAOL-AOT-NRT_V1.3.6_*.hdf
_AOT_LND_Model3	AOT550_mode_3_unfiltered from
	SMAOL-AOT-NRT_V1.3.6_*.hdf
_AOT_LND_Model4	AOT550_mode_4_unfiltered from
	SMAOL-AOT-NRT_V1.3.6_*.hdf
_AOT_LND_Model5	AOT550_mode_5_unfiltered from
	SMAOL-AOT-NRT_V1.3.6_*.hdf
_AOT_LND_Model6	AOT550_mode_6_unfiltered from
	SMAOL-AOT-NRT_V1.3.6_*.hdf
_AOT_LND_Processing_Summary	Processing_Summary from SMAOL-AOT-NRT_V1.3.6_*.hdf
_AOT_LND_Quality_Flags	Quality_Flags from SMAOL-AOT-NRT_V1.3.6_*.hdf
_AOT_LND_Surf_Refl_VIS06	Surf_Refl_VIS06 from SMAOL-AOT-NRT_V1.3.6_*.hdf
OPTIONAL OCEAN DIAGNOSTIC FIELDS	**DESCRIPTION/SOURCE**
_AOT_OCN	Aerosol_Optical_Depth from SEV_AER-OC-L2_*_V1-04.hdf
_AOT_OCN_Angstrom_Exponent	Angstrom_Exponent from SEV_AER-OC-L2_*_V1-04.hdf
_AOT_OCN_DQX	DQX from SEV_AER-OC-L2_*_V1-04.hdf
_AOT_OCN_Model_Number	Model_Number from SEV_AER-OC-L2_*_V1-04.hdf

Source: World Bank.

Note: AOT = aerosol optical thickness; MSG = Meteosat Second Generation; NRT = near real-time; SEVIRI = Spinning Enhanced Visible and Infrared Imager; SMAOL = SEVIRI-MSG Aerosol Over Land.

model identification and the clean AOD product that reports AOD only for high-quality pixels. Also included in the product is the near-clear-sky reflectance at 630 nanometers, quality flags (including land and water and cloud mask identification), and processing information. Each of these fields was extracted for the region of interest. All information was preserved for the region as potential diagnostic information.

An additional field representing an estimate of the nominal AOD uncertainty was generated based on

$$AOT_{UNC} = AOT_{BIAS} + AOT_{RMS} \times SMAOL \qquad (B.2)$$

with values of $AOT_{BIAS} = 0.017$ and $AOT_{RMS} = 0.63$ estimated based on published performance. This formulation does not consider errors due to undetected cloud and other degenerate conditions that might lead to larger errors.

A separate AOD product is available through ICARE over ocean: "SEV_AER-OC-L2_*_V1-04.hdf." Because many regions of interest are located on the coast, the AOD values from the ocean product were optionally included in the extraction process to be used as potential diagnostic information (that is, for nearby coastal pixels). This product includes the AOD, Angstrom Exponent, aerosol model, and quality flags.

The SEVIRI AOD product(s) are reported with a 15-minute refresh for daytime conditions. For the Accra region, SMAOL products were collected between 08Z and 17Z for the period from January 2017 to December 2017. The product was not available through ICARE for only a few days during this period. The ICARE Data and Services Center provided access to the data used in this study.

MERRA CTM data

The MERRA-2 data were downloaded through the Goddard Earth Sciences (GES) Data and Information Services Center (DISC).[4] The MERRA-2 reanalysis product is organized into 26 file collections of instantaneous or time-averaged fields on three-dimensional native, pressure, or edge-level vertical grids or two-dimensional grids of related meteorological, ground surface, chemical, or aerosol variables. To provide inputs to the CTM and statistical approaches, subsets of five data file collections were downloaded in netCDF4 format. The file collections were the inst3_2d_gas_Nx for instantaneous three-hourly aerosol optical depth, tavg1_2d_flx_Nx for time-averaged hourly planetary boundary layer height (PBLH), surface temperature, and surface horizontal wind components (U and V), inst3_3d_aer_Nv for instantaneous three-hourly vertical profile on the 72 model native levels of speciated aerosol, tavg1_2d_aer_Nx for time-averaged hourly speciated surface aerosol and column mass density, and inst3_3d_asm_Nv for surface geopotential and mid-layer height of the 71 model layers. The data file collections and extracted variables from each are summarized in table B.5. The data from all these file collections were processed and combined with satellite optical depths in a Python script and then written to a comma-separated values file for input to the CTM and statistical approach R scripts.

Ground-level monitoring data and land-use variables

2017 GLM data for each city were extracted from the OpenAQ database. Several surface sites had slightly different names in the database because of capitalization and spacing differences, even though they represented data from the same site. Data from these sites were combined before further processing.

TABLE B.5 **MERRA-2 data file collections and extracted variables for input into the chemical transport model and statistical approach scripts**

COLLECTION	VARIABLES	TIME	GRID
inst3_2d_gas_Nx	Aerosol optical depth	Instantaneous, three-hourly	2D
inst3_3d_aer_Nv	Air density	Instantaneous, three-hourly	3D
	Mixing ratio of:		
	Hydrophilic black carbon		
	Hydrophobic black carbon		
	Hydrophilic organic carbon		
	Hydrophobic organic carbon		
	Dust mixing ratio in five size bins		
	Sea salt in five size bins		
	Sulfate aerosol		
Inst3_3d_asm_Nv	Surface geopotential	Instantaneous, three-hourly	3D
	Midlayer height		
tavg1_2d_aer_Nx	Surface mass concentration and column mass density of:	Time-averaged, one-hourly	2D
	Black carbon		
	Organic carbon		
	Sulfate		
	Dust		
	Dust PM$_{2.5}$		
	Sea Salt		
	Sea Salt PM$_{2.5}$		
tavg1_2d_flx_Nx	PBLH	Time-averaged, one-hourly	2D
	Surface temperature		
	Surface U and V		

Source: World Bank.
Note: 2D = two-dimensional; 3D = three-dimensional; MERRA = Modern Era Retrospective–Analysis for Research and Applications; PBLH = planetary boundary layer height; PM$_{2.5}$ = particulate matter with an aerodynamic diameter less than or equal to 2.5 microns.

The GLM data generally included 1-hour average PM$_{2.5}$ data, or 15-minute average data for Delhi. These data were used to calculate daily average PM$_{2.5}$ concentrations. Negative values were presumed to be errors and were removed from the data set. Only days with at least 19 hours of valid data at a given site were used to calculate daily averages.

Land-use variables (including percentage of land that was classified as urban, grassland/wetland/natural, water, barren, or trees; distance to coast; major and minor road length; elevation; and average population density) were calculated for different buffer distances around each GLM network site (100, 200, 500, 750, 100, 1,500, 2,000, 2,500, 3,000, and 3,500 meters) using the data sets for land-use regression in appendix A. Each variable and buffer distance were then used one at a time in a simple linear regression to predict the ground-level PM$_{2.5}$ concentrations. The buffer distance that gave the best overall fit for each variable (measured with the *R* correlation statistic) was used for all further tests of land-use variables.

CTM-based approach

The CTM-based approach used the instantaneous three-hourly MERRA-2 AODs closest to the satellite overpass time and the daily average surface (dry) $PM_{2.5}$ values. This approach assumes that there is a conversion factor of 1.4 to go from organic carbon to organic matter. Ground-level $PM_{2.5}$ values were calculated using the MERRA-2 output for the $PM_{2.5}$ contribution of dust and sea salt aerosols and assuming that all other aerosol components were in the $PM_{2.5}$ fraction. Ground-level PM_{10} was also calculated by including all modeled aerosol species. Other variables, such as the PBL height, the percentage contribution of different aerosol species to the ground-level $PM_{2.5}$, and the percentage of aerosol mass present above the PBL were also calculated based on the MERRA-2 output. The CTM results included for this report did not use the percentage of aerosol mass present above the PBL metric due to computational time constraints, but this screening metric will be included in the final delivered products.

Statistical approach

The statistical approach fitted a GAM using the R software package to predict the ground-level $PM_{2.5}$ concentrations at all the available GLM network sites within a given city. The initial fit used day of week and the satellite AOD divided by the PBL height (AOD/PBLH) as the primary predictors and a log link function.

RESULTS

Local government versus US Diplomatic Post data

A test was performed to determine if the US Diplomatic Post data were roughly consistent with the local GLM network data. The approach used kriging of the local GLM network data to predict the annual average ground-level $PM_{2.5}$ concentration at the US Diplomatic Post and to compare this with the measured annual average at the post. An example of this analysis is shown in the top panel of figure B.2 for Ulaanbaatar. The gridded field produced via kriging the local GLM network data captures the measured value at the US Diplomatic Post well and somewhat captures the large gradient between the post and the Bukhiin Urguu monitoring site. The bottom panel of figure B.2 shows the kriging estimate when the US Diplomatic Post is included. The predictions near the post do not change much, but the combined kriging estimate suggests a more general region of low $PM_{2.5}$ concentrations between Bukhiin Urguu and Amgalan. Based on this analysis, the local GLM network data can be used in combination with the US Diplomatic Post data in the training and evaluation of the satellite $PM_{2.5}$ approaches.

CTM-based approaches versus statistical approaches

The CTM-based and statistical approaches were applied to the MODIS combined product (10-km resolution) to try to predict the daily average $PM_{2.5}$ concentration in Delhi, Lima, and Ulaanbaatar.

FIGURE B.2

Local Ulaanbaatar, Mongolia, ground-level-monitoring network and US Diplomatic Post annual average PM$_{2.5}$ concentrations as well as a gridded estimate produced via ordinary kriging

Micrograms per cubic meter

a. Results without US Diplomatic Post

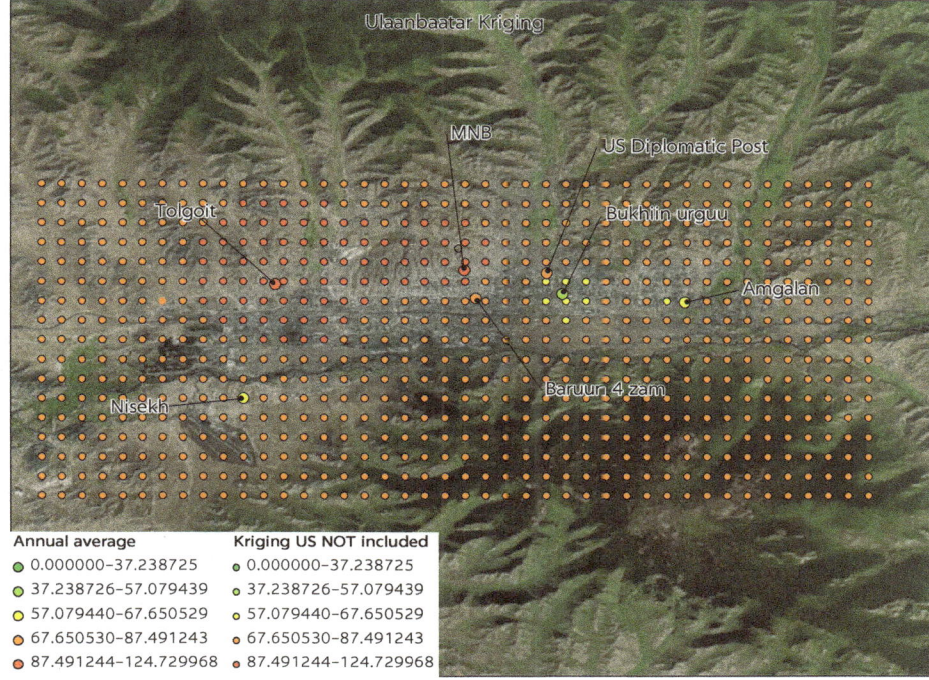

Annual average
- 0.000000–37.238725
- 37.238726–57.079439
- 57.079440–67.650529
- 67.650530–87.491243
- 87.491244–124.729968

Kriging US NOT included
- 0.000000–37.238725
- 37.238726–57.079439
- 57.079440–67.650529
- 67.650530–87.491243
- 87.491244–124.729968

b. Results with US Diplomatic Post

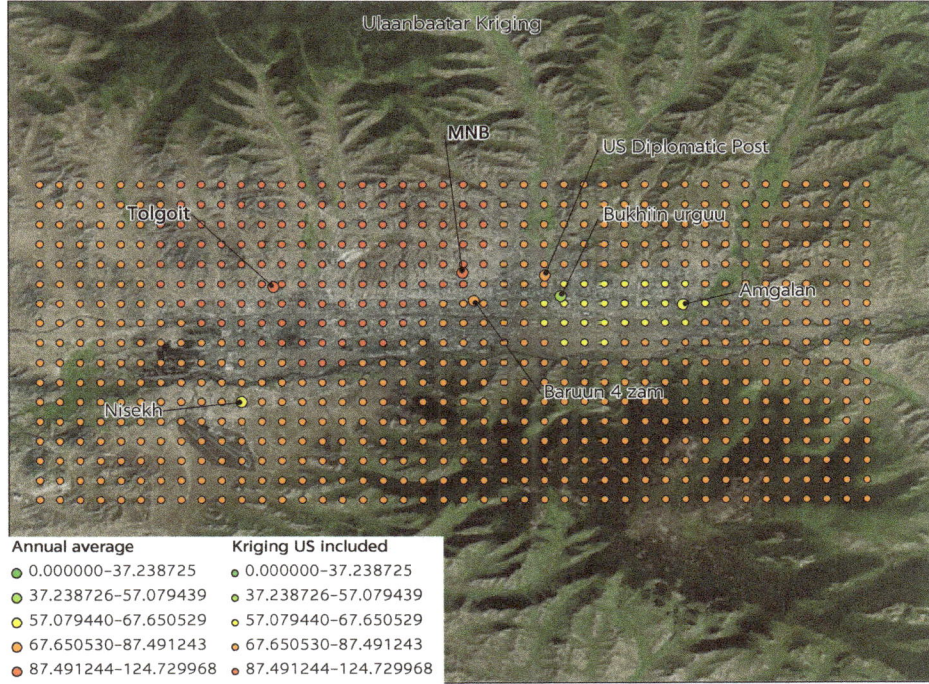

Annual average
- 0.000000–37.238725
- 37.238726–57.079439
- 57.079440–67.650529
- 67.650530–87.491243
- 87.491244–124.729968

Kriging US included
- 0.000000–37.238725
- 37.238726–57.079439
- 57.079440–67.650529
- 67.650530–87.491243
- 87.491244–124.729968

Source: World Bank, produced using Esri ArcGIS.
Note: Spacing between kriging points is 0.01° latitude-longitude (about one kilometer). In panel a, the kriging estimate was produced using local ground-level-monitoring (GLM) network data only. In panel b, the kriging estimate was produced using local GLM data and US Diplomatic Post data. MNB = mean normalized bias; PM$_{2.5}$ = particulate matter with an aerodynamic diameter less than or equal to 2.5 microns.

After filtering for valid AOD values over surface sites from OpenAQ with valid daily average $PM_{2.5}$ measurements, there were no matching observations for the city of Lima, Peru (figure B.3). This is because of the presence of persistent clouds over the region, the fact that this is a coastal city and thus many satellite footprints will have a mix of land and water surfaces, and the relative lack of GLM network data currently in OpenAQ for this city. Thus, Lima could not be evaluated any further in this initial work, but CTM-based ground-level $PM_{2.5}$ estimates were produced for it. Future work will determine if this coverage issue is likely to be a large problem for all coastal cities.

Similarly, in Ulaanbaatar, there were no valid MODIS AOD values for the entire winter (December, January, February, and most of March) because of persistent snow cover. This limits the ability of satellites to represent the true annual average ground-level $PM_{2.5}$ concentrations in this city, as discussed further below.

Statistical approach

In Delhi, the statistical method AOD/PBLH was a statistically significant (at the $\alpha = 0.001$ level), but fairly poor, predictor of daily average ground-level $PM_{2.5}$ values, with R of 0.09 (for MODIS Aqua, 13:30 local standard time overpass) and 0.18 (for MODIS Terra, 13:30 local standard time overpass). Sensitivity studies fitting AOD and PBLH separately slightly increased the correlation (R of 0.21), but in general there was little correlation of MODIS combined product (Deep Blue plus Dark Target) AOD with ground-level $PM_{2.5}$ estimates.

AOD/PBLH was a slightly better predictor in Ulaanbaatar, with R values of 0.30 and 0.18 for MODIS Aqua and Terra, respectively. However, the overall correlation was still fairly poor. In addition, note that there was not a consistent pattern in which the MODIS satellite gave the better correlation with ground-level data, with the afternoon Aqua values performing better in Ulaanbaatar and the morning Terra values performing better in Delhi.

Assessing the GAM fits by examining the residuals showed significant issues. The residuals deviated significantly from a normal distribution and showed a strong dependence on the predicted value—both indications of a poor statistical fit. An example for Ulaanbaatar using MODIS Aqua is shown in figure B.4, which was the best performing statistical fit. Performing simple linear fits of daily average $PM_{2.5}$ and AOD also gave poor correlations, and thus the poor fit is likely not due to the chosen statistical approach (that is, choice of GAM link function), but rather is representative of the low correlation between daily MODIS AOD values and daily average ground-level $PM_{2.5}$ values.

Figure B.5 shows a scatterplot of the observed and GAM-predicted daily average $PM_{2.5}$ values for the Income Tax Office at Delhi (left) and the US Diplomatic Post at Ulaanbaatar (right) for both Aqua (top) and Terra (bottom). The US Diplomatic Post at Delhi was not plotted because a couple of extreme values make it difficult to evaluate the plots. Also note that as only days with a valid AOD value were used, the days plotted for Aqua and Terra were not the same subset of days. The GAM-predicted values tend to have a smaller range of variability than the observed values and exhibit a high level of scatter about the 1:1 line. Tables B.6 and B.7 give the 2017 mean bias (MB), mean normalized bias (MNB), mean normalized gross error (MNGE), and R statistics for all satellite AOD methods tested for Delhi and Ulaanbaatar, respectively. As expected, the statistical approach tends to have a low MB, because the GAM fitting procedure attempts to minimize the bias. However, the MNGE can still be very large

FIGURE B.3

Example of satellite coverage showing the CTM–based estimates for daily average ground-level PM$_{2.5}$ concentrations for all valid MODIS Terra and Aqua aerosol-optical-depth retrievals, Lima, Peru, May 12, 2017

Micrograms per cubic meter

a. MODIS Terra

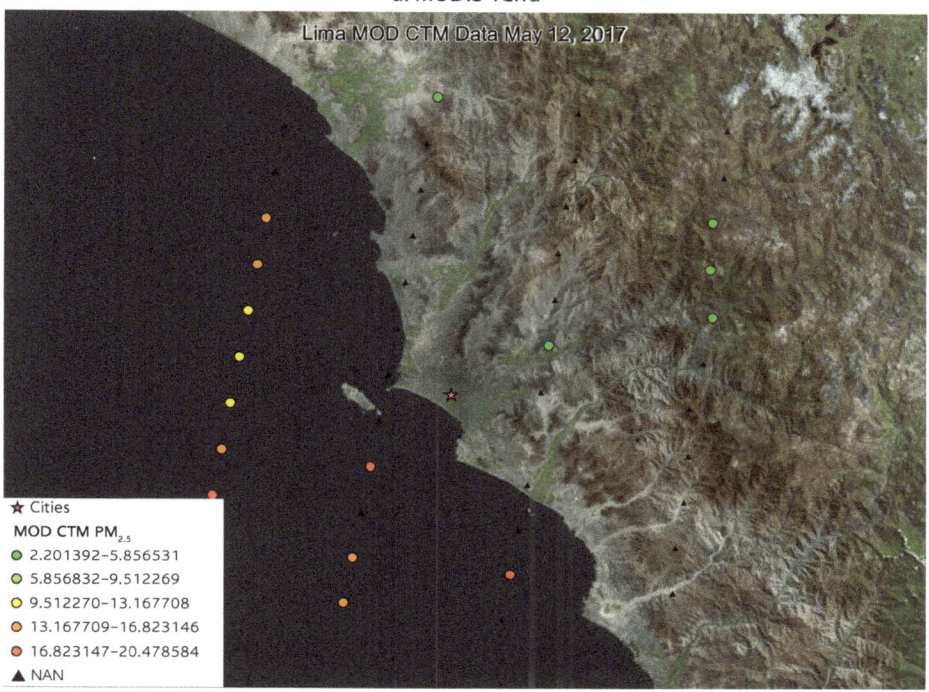

b. MODIS Aqua

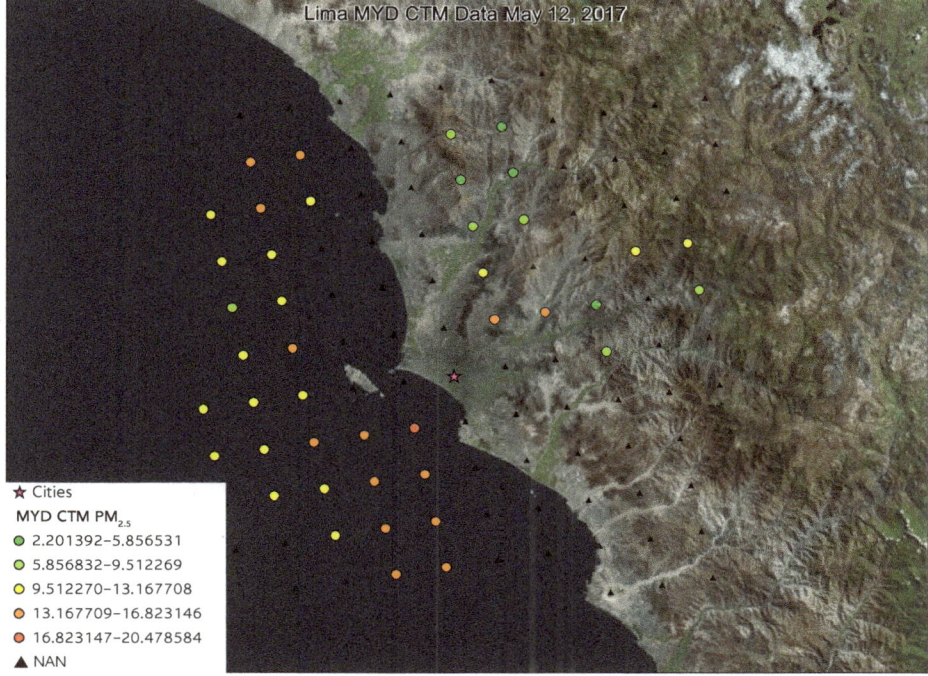

Source: World Bank, produced using Esri ArcGIS.
Note: CTM = chemical transport model; MOD = MODIS Terra; MODIS = Moderate-Resolution Imaging Spectroradiometer; MYD = MODIS Aqua; NAN = not a number; PM$_{2.5}$ = particulate matter with an aerodynamic diameter less than or equal to 2.5 microns.

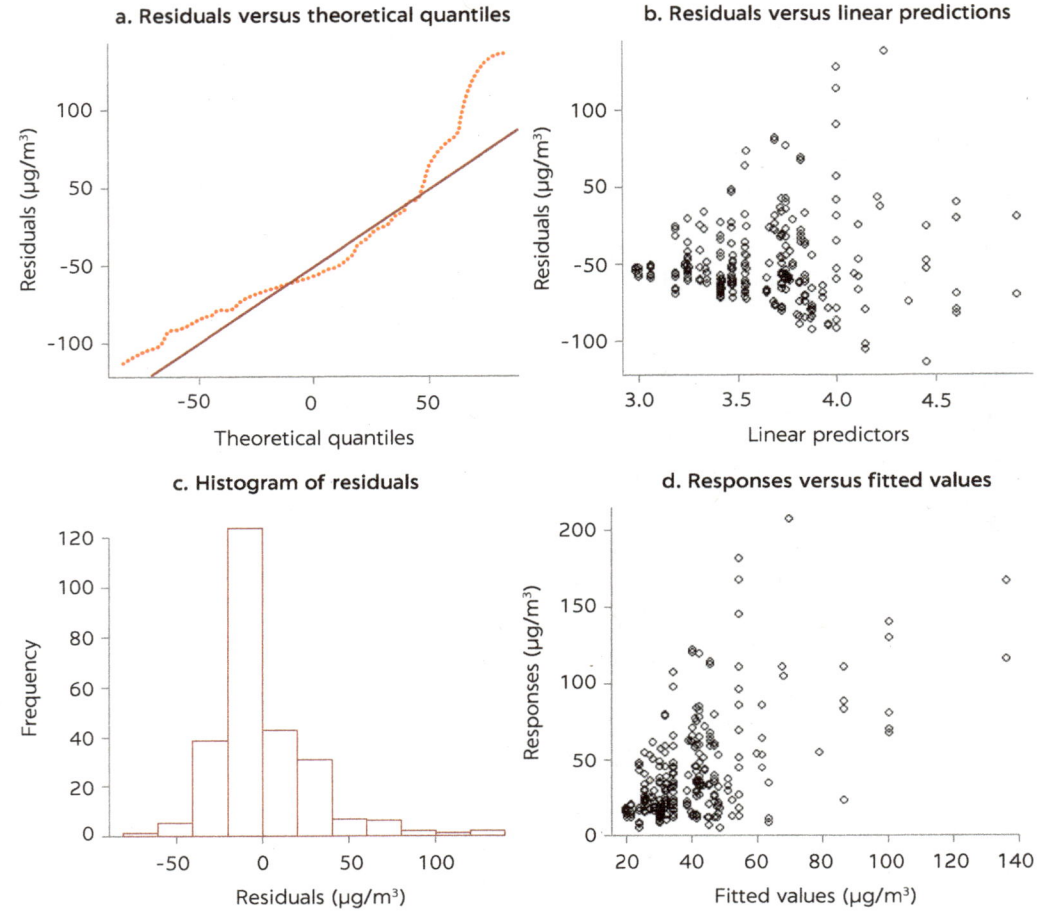

FIGURE B.4

Residual evaluation plots for Ulaanbaatar, Mongolia, using the MODIS Aqua aerosol-optical-depth product

Source: World Bank.
Note: Panel a shows the residual percentiles (*y*) plotted against the predicted quantiles if the residuals were following a normal distribution (*x*). Panel b shows the residuals (*y*) plotted against the value of the linear predictor. The cone shape of the plot suggests a poor statistical fit. Panel c shows the histogram of the residuals. Panel d shows the plot of the generalized additive model response (*y*) versus the measurements (fitted values). MODIS = Moderate-Resolution Imaging Spectroradiometer; $\mu g/m^3$ = micrograms per cubic meter.

TABLE B.6 Statistics for the Delhi, India, satellite ground-level PM$_{2.5}$ products tested in this work

STATISTIC	MODIS TERRA		MODIS AQUA		VIIRS SUOMI-NPP	
	STAT.	CTM	STAT.	CTM	STAT.	CTM
MB (micrograms per cubic meter)	0.045	−0.822	0.113	−29.275	−0.014	−58.102
MNB (%)	30.8	21.7	24.4	−7.5	32.3	−27.3
MNGE (%)	51.5	54.5	44.1	46.8	54.2	52.3
Correlation coefficient (*R*)	0.18	0.17	0.09	0.04	0.11	0.10

Source: World Bank.
Note: "STAT." is the statistical method, and "CTM" is the chemical transport model-based method. MB = mean bias; MNB = mean normalized bias; MNGE = mean normalized gross error; MODIS = Moderate-Resolution Imaging Spectroradiometer; PM$_{2.5}$ = particulate matter with an aerodynamic diameter less than or equal to 2.5 microns; Suomi-NPP = Suomi-National Polar-Orbiting Partnership; VIIRS = Visible Infrared Imaging Radiometer Suite.

FIGURE B.5

Observed and GAM–predicted daily average PM$_{2.5}$ values for the US Diplomatic Posts in Delhi, India, and Ulaanbaatar, Mongolia, for both MODIS Aqua and Terra

Micrograms per cubic meter

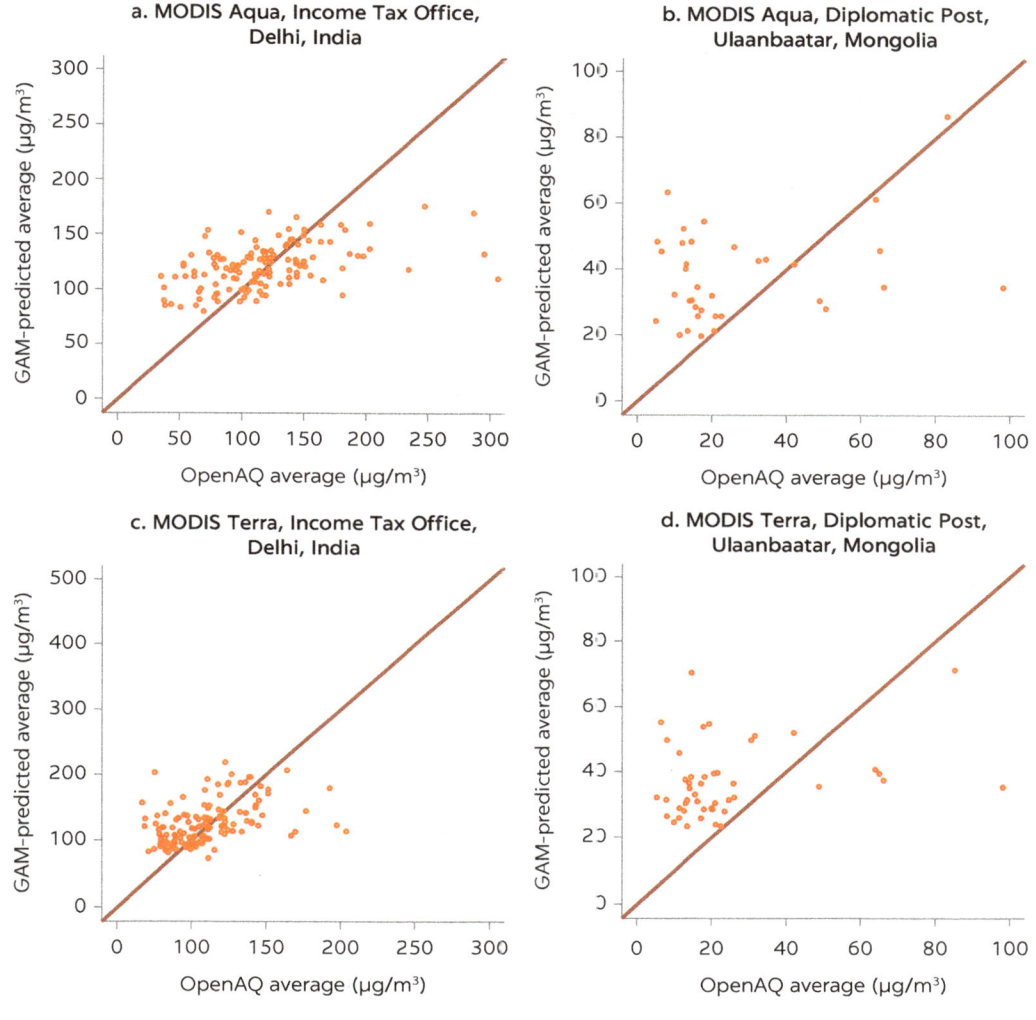

Source: World Bank.
Note: GAM = generalized additive model; MODIS = Moderate-Resolution Imaging Spectroradiometer; PM$_{2.5}$ = particulate matter with an aerodynamic diameter less than or equal to 2.5 microns; µg/m³ = micrograms per cubic meter.

(44–78 percent), again showing that the statistical models have little skill in capturing daily variability or the variability between sites in these cities.

CTM-based approach

Figure B.6 shows a scatterplot of the observed and CTM-based estimates of daily average PM$_{2.5}$ values for the Income Tax Office at Delhi (left) and the US Diplomatic Post at Ulaanbaatar (right) for both the MODIS Aqua (top) and Terra (bottom). The CTM-based estimate for Delhi does a reasonable job representing the mean and variability of the observed values, but individual daily predictions can have significant errors. In contrast, in Ulaanbaatar, the CTM-based estimate

TABLE B.7 **Statistics for the Ulaanbaatar, Mongolia, satellite ground-level PM$_{2.5}$ products tested in this work**

STATISTIC	MODIS TERRA		MODIS AQUA	
	STAT.	CTM	STAT.	CTM
MB (micrograms per cubic meter)	0.082	−31.174	0.443	−33.139
MNB (%)	54.0	−72.2	54.8	−73.1
MNGE (%)	78.5	78.2	78.3	78.8
Correlation coefficient (R)	0.17	0.0	0.30	0.01

Source: World Bank.
Note: "STAT." is the statistical method, and "CTM" is the chemical transport model-based method. MB = mean bias; MNB = mean normalized bias; MNGE = mean normalized gross error; MODIS = Moderate-Resolution Imaging Spectroradiometer; PM$_{2.5}$ = particulate matter with an aerodynamic diameter less than or equal to 2.5 microns.

FIGURE B.6

Observed and CTM–based estimates daily average PM$_{2.5}$ values for the US Diplomatic Posts in Delhi, India, and Ulaanbaatar, Mongolia, for both MODIS Aqua and Terra
Micrograms per cubic meter

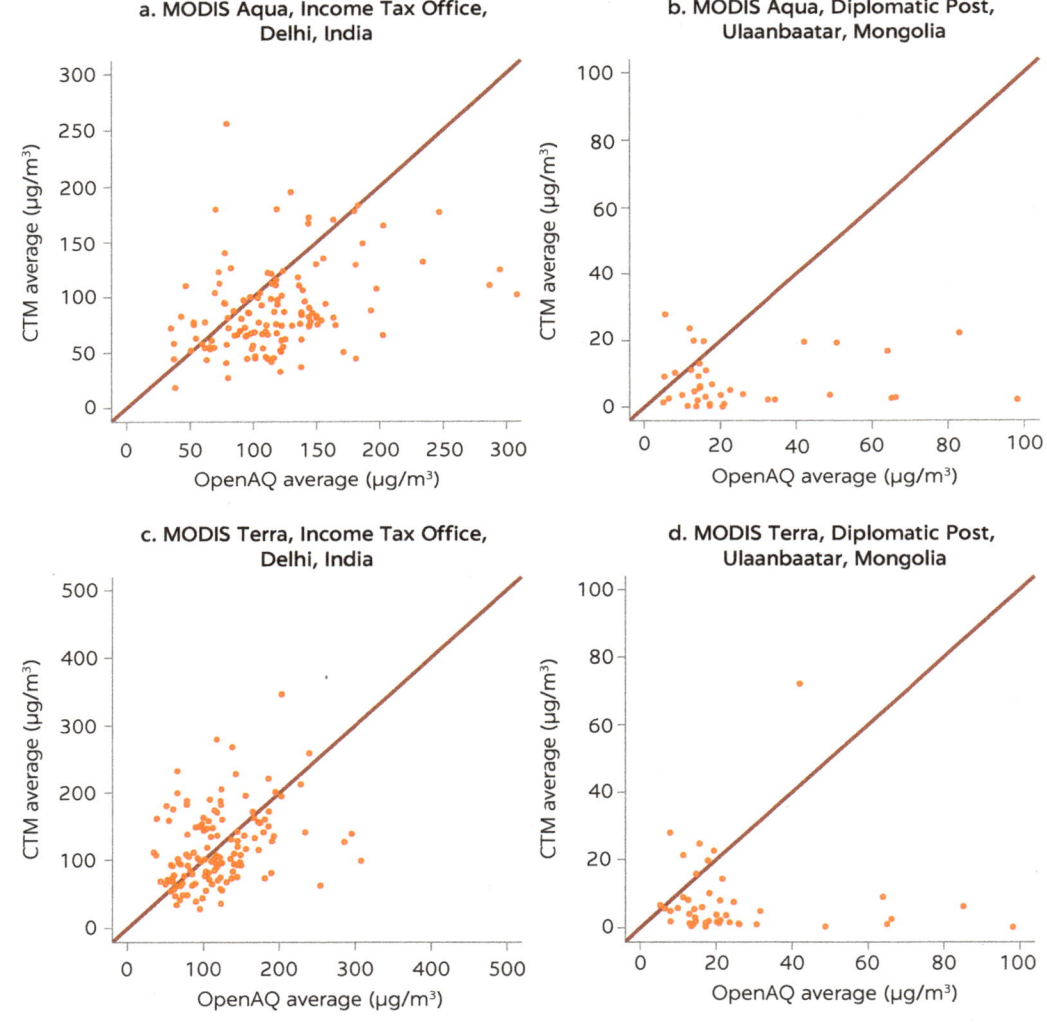

Source: World Bank.
Note: CTM = chemical transport model; MODIS = Moderate-Resolution Imaging Spectroradiometer; OpenAQ = openaq.org; PM$_{2.5}$ = particulate matter with an aerodynamic diameter less than or equal to 2.5 microns; μg/m³ = micrograms per cubic meter.

almost always underestimates the observed values and appears to be doing a poorer job than the statistical approach.

This is consistent with the statistics shown in tables B.6 and B.7. The *R* statistics for the CTM-based approaches using MODIS data tend to be very low (0.01–0.17), which combined with the high MNGE values suggests that the CTM-based method also has little skill in capturing daily variability or the variability between sites in these cities. However, in contrast to the statistical models, MODIS Terra AOD seems to perform better for Ulaanbaatar in the CTM-based approach.

The CTM-based approach has difficulty reproducing daily average values. However, further examination of the CTM-based results for Delhi shows that the scaling of the MERRA-2 daily average PM$_{2.5}$ surface concentrations by the MODIS Terra AOD did result in reasonable values for the annual averages at each surface site. Table B.8 shows the comparison between the observed annual average at each surface site (averaged only over those days with both a valid AOD and a valid daily average PM$_{2.5}$ value, and with the outliers—greater than 600 micrograms per cubic meter—filtered for the US Diplomatic Post), the CTM-based estimate of this annual average, and the annual average predicted by the MERRA-2 model. Using the satellite AOD tends to increase the MERRA-2 ground-level PM$_{2.5}$ concentration by about a factor of two, bringing it much closer to the observed annual averages at each surface site for those days where the surface sites had a valid satellite AOD. However, the low spatial resolution of MODIS AOD product and the MERRA-2 data mean that the CTM-based product has little skill in predicting the variations in the annual average PM$_{2.5}$ values at each site.

TABLE B.8 **Annual average PM$_{2.5}$ surface concentrations for Delhi, India, 2017**

Micrograms per cubic meter

LOCATION	GROUND AVERAGE	CTM-BASED AVERAGE	MERRA-2 AVERAGE
Anand Vihar	153.5	125.8	62.7
Delhi Technological University	123.9	113.5	58.5
Institute of Human Behavior and Allied Sciences	98.9	124.9	63.0
Income Tax Office	119.0	116.9	58.0
Mandir Marg	97.3	111.6	57.4
Netaji Subhas Institute of Technology Dwarka	140.2	112.9	62.4
Punjabi Bagh	103.6	117.0	55.8
Ramakrishna Puram	143.2	122.5	67.9
Shadipur	128.7	117.1	58.1
US Diplomatic Post	113.8	116.3	62.0
Urban average	**122.2**	**117.9**	**60.6**

Source: World Bank.
Note: Annual average surface concentrations are as determined by the GLM data from OpenAQ, the CTM-based satellite approach using MODIS (Moderate-Resolution Imaging Spectroradiometer) Terra AOD, and the original MERRA-2 (Modern Era Retrospective-analysis for Research and Applications) output. These averages include only those days with both a valid AOD and a valid GLM daily average PM$_{2.5}$ value. AOD = aerosol optical depth; CTM = chemical transport model; GLM = ground-level monitoring; PM$_{2.5}$ = particulate matter with an aerodynamic diameter less than or equal to 2.5 microns.

In addition, the uneven satellite AOD coverage over the year could potentially bias the urban annual averages, since some months would be weighted more heavily than others. Thus, the annual average for all GLM network sites in Delhi was calculated as well, not just those with valid satellite AODs. For Delhi, this true GLM average (133.8 micrograms per cubic meter) is 9.5 percent higher than that calculated using only those sites with valid AOD observations, likely because of the poor satellite coverage during the dry season when $PM_{2.5}$ concentrations are expected to peak. This suggests that, even if the conversion of satellite AOD to ground-level $PM_{2.5}$ were perfected for this city, the uneven coverage would bias the satellite-based annual average about 10 percent lower than the true average.

The research conducted for this report tested whether the CTM-based approach in Delhi could be used to estimate monthly averages at a given site. The US Diplomatic Post site was used for this comparison. The monthly average concentrations are shown in figure B.7 for the GLM data, the CTM-based estimate, and the monthly average predicted by the MERRA-2 model. As before, the scaling with the satellite AOD improves upon the MERRA-2 prediction, but neither the MERRA-2 output nor the CTM-based estimate captures the observed seasonal cycle of $PM_{2.5}$ at this location, tending to overestimate during the wet season and underestimate during the dry. Several of these months (December, January, July,

FIGURE B.7

Monthly average $PM_{2.5}$ surface concentration, 2017

Micrograms per cubic meter

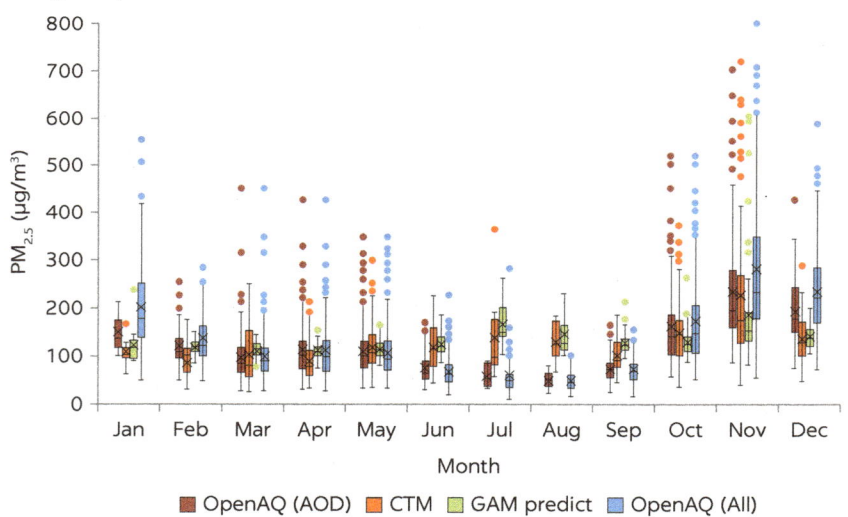

Source: World Bank.
Note: Monthly average $PM_{2.5}$ surface concentration is as determined by the GLM data, the CTM-based satellite approach using MODIS Terra AOD, and the original MERRA-2 output. These averages include only those days with both a valid AOD and a valid GLM daily average $PM_{2.5}$ value. AOD = aerosol optical depth; CTM = chemical transport model; GLM = ground-level monitoring; MERRA = Modern Era Retrospective–analysis for Research and Applications; OpenAQ = openaq.org; $PM_{2.5}$ = particulate matter with an aerodynamic diameter less than or equal to 2.5 microns; μg/m³ = micrograms per cubic meter.

and August) also had fewer than 10 valid matches with the MODIS Terra AOD, which may contribute to the errors. Thus, although the CTM-based approach appears to do a reasonable job for a citywide annual average in Delhi, it does not show any skill in determining the spatial or seasonal variation of PM$_{2.5}$.

The situation is worse for Ulaanbaatar. The use of satellite AOD to scale the MERRA-2 predictions of ground-level PM$_{2.5}$ does improve upon the raw MERRA-2 output for citywide annual averages. However, the initial MERRA-2 estimate is so low (6.1 micrograms per cubic meter) that the scaled value (7.1 micrograms per cubic meter) is still a dramatic underprediction of the observed value (40.3 micrograms per cubic meter) for days when there is a valid satellite AOD. This suggests that in Ulaanbaatar, the relationship between ground-level PM$_{2.5}$ and the AOD is so different from the MERRA-2 estimate that the CTM-based approach does not appear to add any value to the existing GLM data set. Thus, using the CTM-based approach in Ulaanbaatar would require running a CTM that gives a better initial estimate of the vertical profile and optical properties of the aerosol in this region. Further progress in using satellite observations in Ulaanbaatar would likely benefit from adding a SPARTAN network site to the city.

However, due to the lack of satellite AOD during the winter months, the true annual average PM$_{2.5}$ concentration (88.1 micrograms per cubic meter) is much higher than that calculated when only the ground sites with valid AOD observations are used (40.3 micrograms per cubic meter). This suggests that, even if the conversion of satellite AOD to ground-level PM$_{2.5}$ were perfected for Ulaanbaatar, the lack of satellite AOD values in winter would bias the satellite-based annual average about 50 percent lower than the true average.

Comparison with GBD 2016 estimates

In Delhi, the Global Burden of Disease (GBD) 2016 data set[5] gives 2016 annual averages of about 135 micrograms per cubic meter at both 0.1° by 0.1° resolution and 0.01° by 0.01° resolution after GWR bias correction. However, the GBD 2016 estimate before GWR bias correction is 105 micrograms per cubic meter, consistent with the 2017 CTM-based estimate for Delhi of 112 micrograms per cubic meter, and with the explanation of how the relative lack of valid satellite AOD data during December–January could bias the satellite-derived annual average low for these cities.

In Ulaanbaatar, the GBD 2016 data set interestingly gives 2016 annual averages of 30 micrograms per cubic meter at 0.1° by 0.1° resolution and 42 micrograms per cubic meter at 0.01° by 0.01° resolution (after GWR bias correction). These results are pretty close to the "best possible" satellite estimate of 40.3 micrograms per cubic meter, given the lack of wintertime AOD data. However, the GBD 2016 estimate before GWR bias correction is 7 micrograms per cubic meter, consistent with the 2017 CTM-based estimate for this city. Thus, it appears that (1) the GBD 2016 data set also has difficulty representing the annual average in this city, (2) most of the improvement in Ulaanbaatar in GBD 2016 is coming from the global GWR bias correction, and (3) even with this correction, the annual average is underestimated by about a factor of two. Thus, both the GBD 2016 results and our results suggest that GLM is required to get an accurate annual average PM$_{2.5}$ concentration in Ulaanbaatar.

Comparing satellite AOD data sets

VIIRS versus MODIS in Delhi, India

Using the VIIRS AOD retrievals instead of the MODIS products in both the CTM-based and statistical approaches was attempted for Delhi. In the statistical approach, the quality of fit using the VIIRS AOD was not appreciably better than that using MODIS Aqua (which has a similar afternoon equator crossing time). With VIIRS the R was 0.1, as opposed to 0.08 for MODIS Aqua, and there was similar evidence of a poor statistical fit. The scatterplot of the GAM predictions versus the observed daily average values at the Income Tax Office site (panel a of figure B.8) also does not indicate any improved performance relative to using MODIS Aqua or Terra data. The MNB and MNGE are slightly worse when VIIRS is used instead of MODIS Aqua (table B.6).

Using the CTM-based approach with VIIRS AOD data does result in greater spatial coverage over a city (figure B.9), but it results in poorer performance relative to using MODIS (panel b of figure B.8). When the VIIRS data are used, the CTM approach tends to underestimate the ground-level $PM_{2.5}$ concentration, leading to an underestimate of about a factor of two of the annual average ground-level $PM_{2.5}$ concentration (that is, VIIRS-based prediction of 74.8 micrograms per cubic meter versus an observed value of 140.3 micrograms per cubic meter). Thus, unlike the case when the MODIS Terra AOD was used, using the afternoon VIIRS AOD does not significantly improve the MERRA-2 predicted annual average (62.9 micrograms per cubic meter). This is reflected in the large mean bias for this approach in table B.6.

MODIS versus SEVIRI in Accra, Ghana

The predictions of the CTM-based method using the geostationary SEVIRI instrument were compared with those from the polar-orbiting MODIS

FIGURE B.8

Observed values and VIIRS-based estimates of daily average $PM_{2.5}$ concentrations for the Income Tax Office in Delhi, India, using GAM and CTM–based approaches

Micrograms per cubic meter

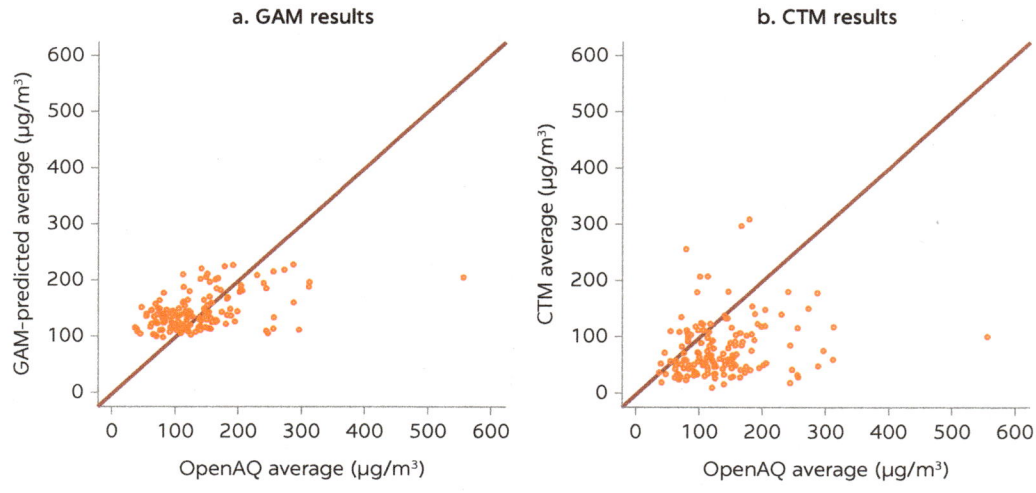

Source: World Bank.
Note: CTM = chemical transport model; GAM = generalized additive model; OpenAQ = openaq.org; $PM_{2.5}$ = particulate matter with an aerodynamic diameter less than or equal to 2.5 microns; µg/m³ = micrograms per cubic meter; VIIRS = Visible Infrared Imaging Radiometer Suite.

FIGURE B.9

A comparison of the CTM–based estimates for daily-average ground-level PM$_{2.5}$ concentrations over Delhi, India, using VIIRS and MODIS Terra, November 1, 2017

Micrograms per cubic meter

a. Using VIIRS

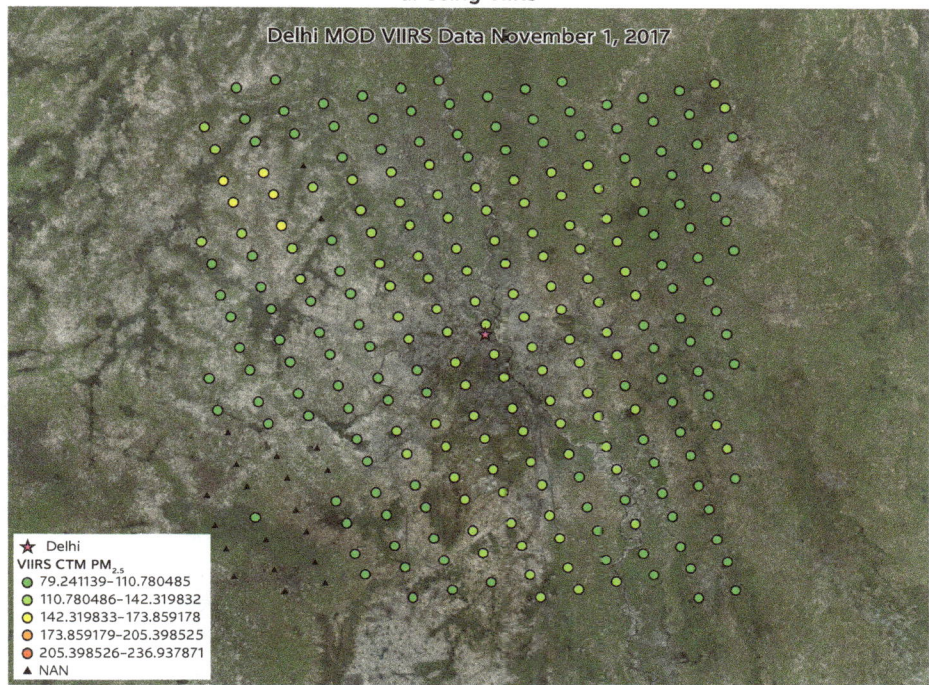

b. Using MODIS Terra

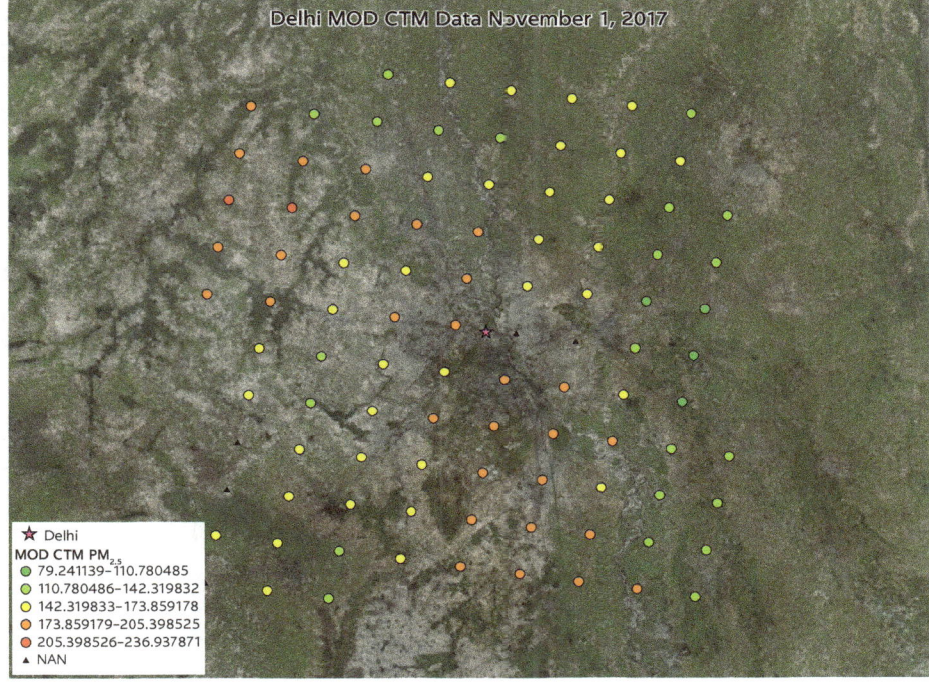

Source: World Bank, produced using Esri ArcGIS.

Note: CTM = chemical transport model; MOD = MODIS Terra; MODIS = Moderate-Resolution Imaging Spectroradiometer; NAN = not a number; PM$_{2.5}$ = particulate matter with an aerodynamic diameter less than or equal to 2.5 microns; VIIRS = Visible Infrared Imaging Radiometer Suite.

FIGURE B.10

A comparison of the CTM–based estimates for daily-average ground-level PM$_{2.5}$ concentrations over Accra, Ghana, using SEVIRI and MODIS Aqua, December 26, 2017

Micrograms per cubic meter

a. SEVIRI data

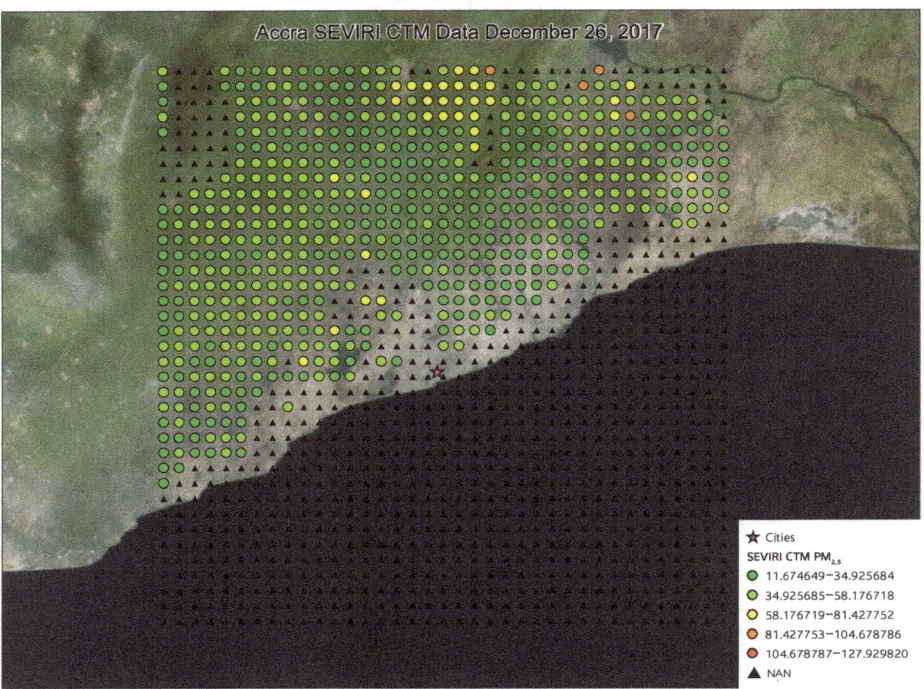

b. MODIS Aqua data

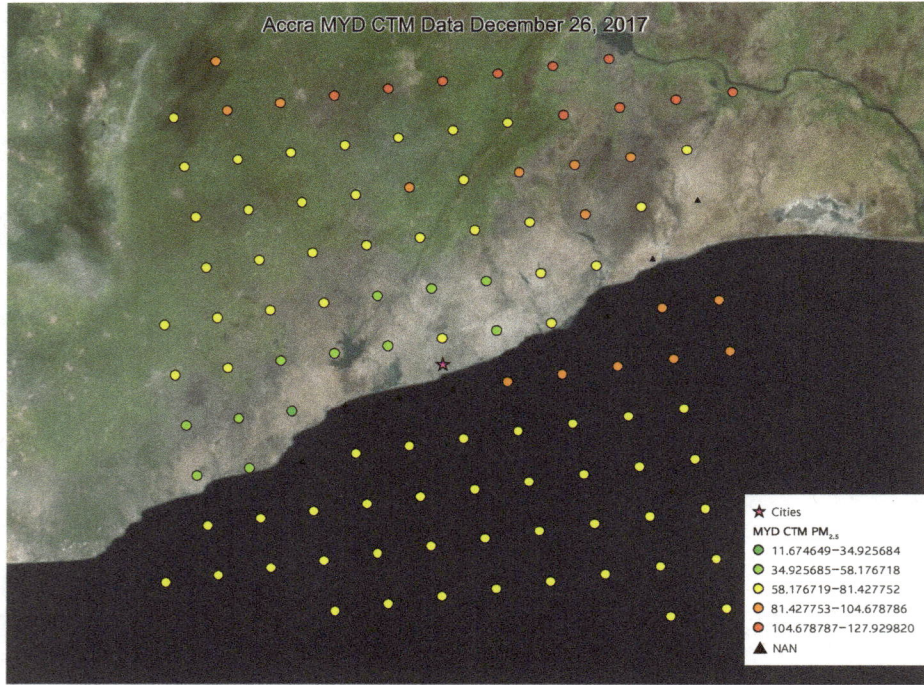

Source: World Bank, produced using Esri ArcGIS.

Note: CTM = chemical transport model; MODIS = Moderate-Resolution Imaging Spectroradiometer; MYD = MODIS Aqua; NAN = not a number; PM$_{2.5}$ = particulate matter with an aerodynamic diameter less than or equal to 2.5 microns; SEVIRI = Spinning Enhanced Visible and Infrared Imager; VIIRS = Visible Infrared Imaging Radiometer Suite.

instrument, using the coastal city of Accra as the test case. The results are shown in figure B.10 for a single day in December. The SEVIRI product has a much higher spatial resolution, as expected, but also has less coverage near the coast. Thus, the SEVIRI product has limited coverage near the city of Accra itself, although it does have reasonable coverage for the area further inland. In contrast, the MODIS combined (Deep Blue and Dark Target) AOD product has much better coverage for this period, although the over-ocean CTM-based values appear higher than expected. The low coastal coverage of the SEVIRI product suggests that it likely would not improve the ability of satellites to monitor PM$_{2.5}$ in Accra over using the MODIS product.

Bias corrections of the CTM-based estimates

A GWR correction to the Stage 1 CTM-based ground-level PM$_{2.5}$ estimates in Delhi was attempted with the *spgwr* software package in the R statistical software platform using only the in-city data. A GWR is like a simple linear regression except the coefficients are allowed to vary with latitude and longitude. The first attempt tried to fit the daily bias in the Stage 1 estimates to land-use variables (percentage of urban land cover, elevation, and population density) using a simple linear regression model. This showed that these variables have little ability to predict the bias in the Stage 1 estimates (R less than 0.01). However, the MERRA-2 predicted values of the concentration (micrograms per cubic meter) of ground-level PM$_{2.5}$ due to organic carbon, sulfate, dust, and sea salt had some skill in predicting the bias, with an R value of 0.37 for a simple linear model fit, but implementing this regression equation as a GWR model instead does not appreciably improve the model fit (R of 0.37). This is likely because Delhi has only 10 GLM sites, as opposed to the 1,440 sites used in van Donkelaar and others (2015), and thus allowing the coefficients to vary with location does not provide much of a benefit. Allowing the regression coefficients to vary with location likely will not provide much of a benefit on a city scale, but this approach could be useful when using high-quality, harmonized GLM measurement networks on country or continent scales, such as exist in North America (van Donkelaar and others 2015) and western Europe (Vienneau and others 2013).

Using the simple linear model to predict the bias and applying this bias correction did improve the correlation of the daily average CTM-based satellite predictions (R of 0.23 compared to 0.15 before bias correction). However, this simple bias correction results in several data points predicted to have negative values, and thus the bias-corrected value produced a much lower citywide annual average (85.4 micrograms per cubic meter) than the initial CTM-based estimate.

Training a simple linear model to predict the actual PM$_{2.5}$ daily average concentration using the above MERRA-2 variables and the Stage 1 CTM-based estimate was then attempted. This did remove the model mean bias for the urban-scale annual average and gave an R value of 0.33, an MNB of 20 percent, and an MNGE of 38 percent. However, the Stage 1 CTM-based estimate was not a significant predictor in this model, and a similar fit can be found by removing the satellite AOD-derived estimate entirely and just fitting the MERRA-2–speciated aerosol variables.

Similarly, applying a bias correction to the Ulaanbaatar data does fix the mean bias in the citywide annual average, but the CTM-based estimate was still unable to predict daily and site-to-site variability (R of 0.07). Thus, the statistical bias corrections of the CTM-based estimates attempted here do not significantly improve the ability of the satellite data to represent the daily and site-to-site variability of $PM_{2.5}$ in either Delhi or Ulaanbaatar.

CONCLUSIONS AND RECOMMENDATIONS

Several novel methods for using satellite AOD from publicly available sources to predict observed ground-level $PM_{2.5}$ daily averages in Lima, Delhi, Ulaanbaatar, and Accra, for 2017 were tested. The research conducted for this report found several limitations in the use of satellite AOD in these cities. For example, all cities had significant limitations in satellite AOD coverage due to persistent clouds, wintertime snow cover, or mixed water-land and other bright surfaces. Satellite coverage was poorest in the coastal city of Lima, and thus a meaningful evaluation of a CTM-based method or the creation of a statistical method was not possible for this city. In Ulaanbaatar, no satellite observations are available for the high $PM_{2.5}$ winter months of December to mid-March, and thus even a perfect method for converting satellite AOD to ground-level $PM_{2.5}$ would underestimate the true annual average $PM_{2.5}$ concentration for this city by 50 percent. Delhi has observations in all months, but substantially fewer observations in the peaks of the wet and dry seasons (December, January, July, and August), which would also result in a slight (about 10 percent) underestimate of the true annual average.

A CTM-based approach could provide a reasonable citywide annual average in Delhi, but the CTM-based approach leads to dramatic underestimates of the citywide annual average in Ulaanbaatar. Similar results were seen in the GBD 2016 data set before GWR bias correction. This is likely due to Ulaanbaatar being in a river valley surrounded by mountains and mostly rural land. Thus, the coarse resolution of global CTMs (combined with likely inaccurate emission inventories for Ulaanbaatar) means that CTMs are not able to correctly represent the aerosol profile within the city, instead using an average profile more representative of the surrounding rural area.

However, neither the CTM-based approach nor the statistical approach tested here were able to represent more than 30 percent of the variability in the daily average $PM_{2.5}$ values within a city over the year, and many techniques had little correlation with the corresponding daily averages. Site-specific monthly average values were more reasonable in Delhi, but the lack of data during several months means that the satellites did a poor job of representing the observed seasonal cycle of $PM_{2.5}$.

Using VIIRS instead of MODIS did not appreciably change the overall evaluation results in Delhi, but the VIIRS product does provide a slightly higher spatial resolution than MODIS (6-kilometer versus 10-kilometer). Although the SEVIRI AOD product does provide a higher spatial resolution, it appears to have lower coverage for coastal cities such as Accra.

The following recommendations are made for LMICs considering the use of satellite AOD in their $PM_{2.5}$ monitoring:

- The physical limits of satellite AOD coverage will make their use in $PM_{2.5}$ monitoring difficult in many cities.
  - In cities with persistent wintertime snow cover, like Ulaanbaatar, it is likely not possible to get accurate annual averages from current satellites.

- In coastal cities, the mix of water and land surfaces within a satellite footprint and persistent clouds may also mean that using satellite AOD is not an option.
- CTM-based methods are also likely to fail for cities that are in appreciably different air-quality environments than their surroundings, such as cities in mountain valleys surrounded by rural land. This means that such cities in Tier I and II countries (that lack the GLM data for statistical approaches) will likely not be able to use satellites in their PM$_{2.5}$ monitoring.
- Estimating variations in annual average PM$_{2.5}$ within a city (to estimate chronic health effects) is unlikely to be possible with satellite AOD data using the approaches tested here. This will require GLM network data and land-use regression, with the satellite AOD product or the satellite-based PM$_{2.5}$ estimate used as a predictor variable in the LUR.
- Both the CTM-based and statistical approaches tested here showed little ability to represent the day-to-day variability in PM$_{2.5}$ concentrations, with average errors (MNGE) of ±50 percent for the best approaches within each city. Thus, studies of acute health effects will likely require GLM data.
- For cities interested in using satellite AOD in their PM$_{2.5}$ monitoring, establishing at least one SPARTAN network site within the city to directly measure the relationship between AOD and ground-level PM$_{2.5}$ will likely help to reduce the errors in the approaches tested here.

NOTES

1. https://ladsweb.modaps.eosdis.nasa.gov/archive/allData/6/MOD04_L2.
2. https://www.avl.class.noaa.gov/saa/products/welcome.
3. http://www.icare.univ-lille1.fr.
4. https://disc.gsfc.nasa.gov/.
5. Downloaded from http://fizz.phys.dal.ca/~atmos/martin/?page_id=140.

REFERENCES

Chatfield, R., M. Sorek-Hamer, A. Lyapustin, and Y. Liu. 2017. "Daily Kilometer-Scale MODIS Satellite Maps of PM$_{2.5}$ Describe Wintertime Episodes." Paper 263193 presented at the "Air and Waste Management Association 2017 Annual Conference and Exhibition," Pittsburgh, PA, June 5–9.

GBD (Global Burden of Disease) 2016 Risk Factors Collaborators. 2017. "Global, Regional, and National Comparative Risk Assessment of 84 Behavioural, Environmental and Occupational, and Metabolic Risks or Clusters of Risks, 1990–2016: A Systematic Analysis for the Global Burden of Disease Study 2016." *Lancet* 390: 1345–422.

Remer, L. A., S. Mattoo, R. C. Levy, and L. A. Munchak. 2013. "MODIS 3 km Aerosol Product: Algorithm and Global Perspective." *Atmospheric Measurement Techniques* 6: 1829–44.

Sorek-Hamer, M., A. W. Strawa, R. B. Chatfield, R. Esswein, A. Cohen, and D. M. Broday. 2013. "Improved Retrieval of PM$_{2.5}$ from Satellite Data Products Using Non-Linear Methods." *Environmental Pollution* 182: 417–23.

van Donkelaar, A., R. V. Martin, M. Brauer, R. Kahn, R. Levy, C. Verduzco, and P. J. Villeneuve. 2010. Global Estimates of Ambient Fine Particulate Matter Concentrations from Satellite-Based Aerosol Optical Depth: Development and Application." *Environmental Health Perspectives* 118 (6): 847–55.

van Donkelaar, A., R. V. Martin, R. J. Spurr, and R. T. Burnett. 2015. "High-Resolution Satellite-Derived PM$_{2.5}$ from Optimal Estimation and Geographically Weighted Regression over North America." *Environmental Science & Technology* 49 (17): 10482–91.

Vienneau, D., K. de Hoogh, M. J. Bechle, R. Beelen, A. van Donkelaar, R. V. Martin, D. B. Millet, G. Hoek, and J. D. Marshall. 2013. "Western European Land Use Regression Incorporating Satellite- and Ground-Based Measurements of NO$_2$ and PM$_{10}$." *Environmental Science & Technology* 47 (23): 13555–64.

Evaluation of Satellite Approaches

The methods (discussed in appendix B) for using satellite observations to determine ground-level particulate matter with an aerodynamic diameter less than or equal to 2.5 microns ($PM_{2.5}$) were applied to nine selected low- and middle-income country (LMIC) cities (table C.1) that cover a mixture of regimes (coastal versus inland, high versus low altitude, tropical versus temperate, and so forth). These cities also have available the ground-level-monitoring (GLM) data needed for validation, through either US Diplomatic Posts, local monitoring networks, or both. This appendix presents the results of the evaluation of the satellite approaches for each city in table C.1.

A 1° by 1° latitude-longitude swath of Moderate-Resolution Imaging Spectroradiometer (MODIS) afternoon Aqua (MYD) and morning Terra (MOD) data was downloaded for each city, centered on the coordinates listed in table C.2. Modern Era Retrospective-analysis for Research and Applications (MERRA) data were also downloaded for each site. For $PM_{2.5}$ monitor data, OpenAQ data from 2016 and 2017 were used and processed for the cities. In addition, individual $PM_{2.5}$ data sets for Accra, Ghana, and Dakar, Senegal, were provided respectively by the Environmental Protection Agency of the Ghana Ministry of Environment, Science, Technology and Innovation (EPA Ghana) and the Center for Air Quality Management, a part of the Senegal Ministry of the Environment and Sustainable Development. The analyses of the data sets for Accra and Dakar focused on the years 2015 and 2011, respectively, since these years had the most-recent fairly continuous $PM_{2.5}$ data for each city. As noted in table C.2, not all cities had GLM sites, and some had only a few.

For each city, both the statistical and chemical-transport-model (CTM)–based approaches described in appendix B were attempted to determine ground-level $PM_{2.5}$ concentrations. Some cities had too few GLM observations to allow a statistical approach to be performed; for these, only the CTM-based approach was evaluated. The statistical and CTM-based $PM_{2.5}$ estimates were then compared with the GLM values for the cases where a valid satellite aerosol optical depth (AOD) existed for the GLM site, and a statistical and graphical evaluation of the match between the satellite estimates and the GLM observations was performed.

TABLE C.1 **Cities included in this study**

CITY	COUNTRY	LOCATION	INCOME GROUP
Accra	Ghana	Coastal, low altitude	Lower middle
Addis Ababa	Ethiopia	Inland, high altitude	Low
Dakar	Senegal	Coastal, low altitude	Low
Delhi	India	Inland, low altitude	Lower middle
Hanoi	Vietnam	Inland, low altitude	Lower middle
Kampala	Uganda	Inland but near lake, high altitude	Low
Kathmandu	Nepal	Inland, high altitude	Low
Lima	Peru	Coastal, low altitude	Upper middle
Ulaanbaatar	Mongolia	Inland, high altitude	Lower middle

Source: World Bank.

TABLE C.2 **Coordinates and number of ground-level monitoring sites, by city**

CITY	COUNTRY	LATITUDE (°)	LONGITUDE (°)	SITES
Accra	Ghana	5.55°	−0.2°	1
Addis Ababa	Ethiopia	9.03	38.74	2
Dakar	Senegal	14.6928	−17.4467	4[a]
Delhi	India	28.61	77.23	10
Hanoi	Vietnam	21.0283	105.854	1
Kampala	Uganda	0.313611	32.5811	1
Kathmandu	Nepal	27.7114	85.3086	4
Lima	Peru	−12.0433	−77.0283	10
Ulaanbaatar	Mongolia	47.9167	106.917	8

Source: World Bank.
a. Dakar data are available only as an average of the four reporting sites.

In addition, for cities with hourly $PM_{2.5}$ data from the GLM monitoring performed at US Diplomatic Posts, the average ground-level $PM_{2.5}$ concentrations during the two-hour window around the satellite overpass time (about 10:30 and 13:30 local solar time at the equator for Terra and Aqua, respectively) were compared to the daily average concentrations. This was done to estimate how well a "perfect" polar-orbiting satellite, which is able to observe a location only once per day, would be able to determine the daily average $PM_{2.5}$ concentration in a city. This evaluation, which requires only GLM data, provides an upper limit on the ability of polar-orbiting satellites to determine ground-level $PM_{2.5}$ concentrations in each city and can be used by LMICs to assess whether satellite data will be appropriate for their monitoring plans.

The second section contains the evaluations for each city. In addition, the third section has data on the changes that can be introduced by applying the OpenAQ-developed Quality Filter to the OpenAQ GLM data. The last section describes an experiment to see if the use of satellite data could allow for a reduction in the number of GLM sites needed to characterize ground-level $PM_{2.5}$ concentrations and exposure within a city, using Delhi, India, and Ulaanbaatar, Mongolia, as test cases.

EVALUATION FOR EACH CITY

Delhi, India

Delhi, India, was one of the few cities that had enough data to train a statistical (generalized additive model or GAM; see appendix B) model with both Aqua and Terra satellite data. Figure C.1 shows the evaluation plots for the residuals of the GAM trained on the Terra data. The response residuals do not follow a normal distribution, which indicates that the statistical fit is fairly poor. Figure C.2 shows that the statistical estimate of ground-level $PM_{2.5}$ increases monotonically with the ratio of the satellite aerosol optical depth (AOD) to the planetary boundary layer (PBL) height, as expected, but the slope of the fit appears to be dominated by two high values of AOD / boundary layer height (PBLH).

Figure C.3 shows a scatterplot of the ground-level $PM_{2.5}$ estimate from the CTM-based method and the statistical method, respectively, against the OpenAQ data for the US Diplomatic Post monitor site. The other OpenAQ sites have similar scatterplots (not shown). Table C.3 shows the evaluation statistics from both

FIGURE C.1

Residual evaluation plots for the generalized additive model trained on the Terra satellite data, Delhi, India

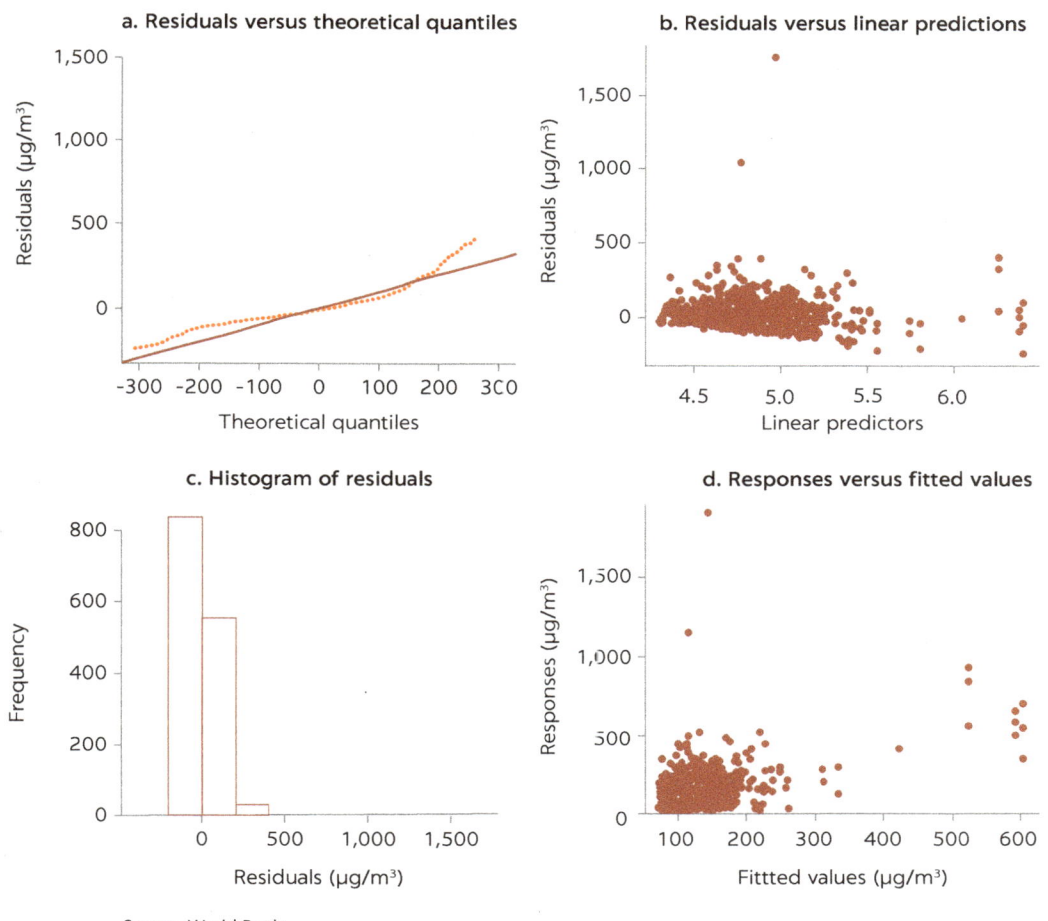

Source: World Bank.
Note: µg/m³ = micrograms per cubic meter.

FIGURE C.2

Plot of the dependence of the ground-level PM$_{2.5}$ estimate from the generalized additive model on the ratio of the Terra aerosol optical depth and the MERRA planetary-boundary-layer height, Delhi, India

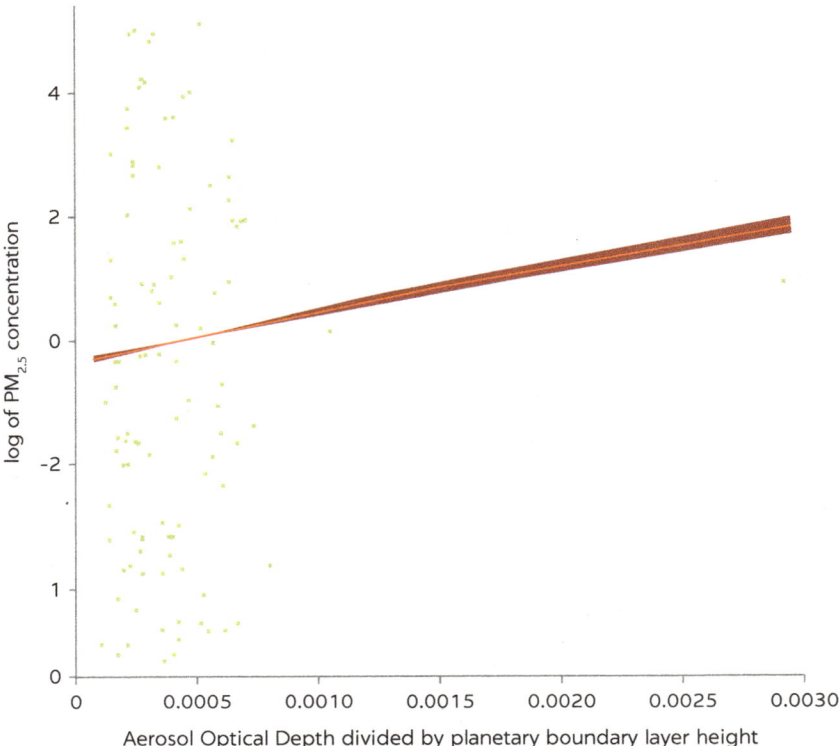

Source: World Bank.
Note: Green dots = individual observations; red bar = uncertainty of the fit; MERRA = Modern Era Retrospective-analysis for Research and Applications; PM$_{2.5}$ = particulate matter with an aerodynamic diameter less than or equal to 2.5 microns.

FIGURE C.3

Scatterplots of the Terra satellite estimate of PM$_{2.5}$ from the generalized additive model and the chemical transport model versus the OpenAQ data, US Diplomatic Post in Delhi, India

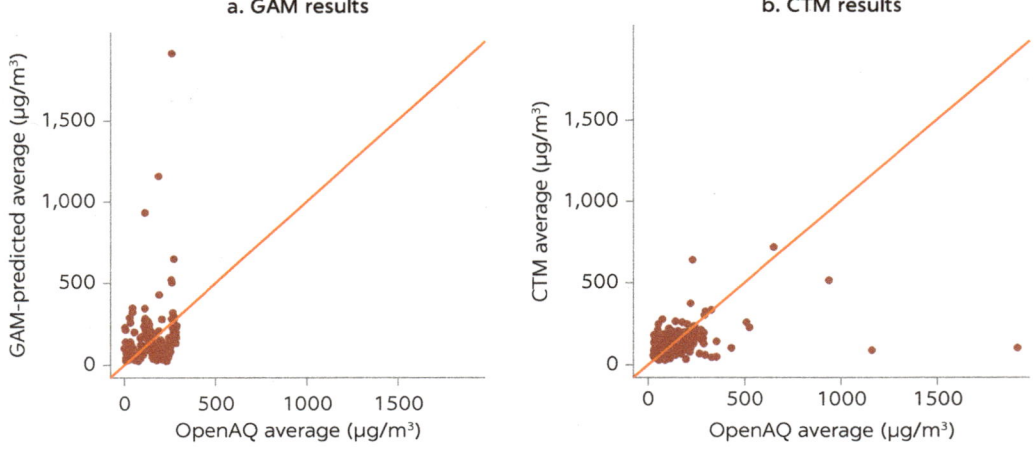

Source: World Bank.
Note: CTM = chemical transport model; GAM = generalized additive model; OpenAQ = openaq.org; PM$_{2.5}$ = particulate matter with an aerodynamic diameter less than or equal to 2.5 microns; µg/m³ = micrograms per cubic meter.

TABLE C.3 Delhi, India, evaluation statistics for the generalized-additive-model- and chemical-transport-model-based methods

	TERRA		AQUA	
STATISTIC	GAM	CTM	GAM	CTM
Correlation coefficient (R)	0.456	0.437	0.328	0.273
Mean bias (micrograme percubic meter)	−0.475	−0.884	0.079	−30.267
Mean normalized bias (%)	36.9	26.9	29.4	−6.1
Mean normalized gross error (%)	59.8	60.4	50.1	49.3
Root-mean-square error (micrograms per cubic meter)	89.4	98.3	83.5	95.5

Source: World Bank.
Note: CTM = chemical transport model; GAM = generalized additive model.

the statistical and CTM-based methods with respect to the OpenAQ data. The correlation coefficients are fairly low (maximum R of 0.46) but the statistical and CTM-based methods using the Terra data have low mean biases (less than 1 microgram per cubic meter).

Comparing these correlation coefficients with the correlation coefficients between the GLM $PM_{2.5}$ concentrations averaged around the satellite over-pass time (two-hour window) and the daily average $PM_{2.5}$ concentrations provides an upper limit on the correlation between the once-daily polar-orbiting satellite observations and the GLM data. For Delhi, these correlations are 55 percent at the Terra overpass time and 64 percent at the Aqua overpass time. Interestingly, these are the opposite of the results in table C.3, which imply that Terra AOD has a better correlation than Aqua AOD. The current performance of the Terra AOD (R of 44–46 percent for the daily average $PM_{2.5}$) is fairly close to the "upper limit" of 55 percent expected for this satellite based on the GLM data correlations.

Figure C.4 shows the monthly averages calculated using the CTM-based and statistical methods and compares them to the values calculated using the OpenAQ data, either using all the available data (OpenAQ All) or only the data on days with a valid AOD retrieval (OpenAQ AOD). During the summer months (wet season), both the CTM-based and statistical methods overestimate the $PM_{2.5}$ values, and they tend to underestimate in the winter months (dry season). Figure C.5 shows the annual average $PM_{2.5}$ estimated for each method by monitoring site. As noted in appendix B, the satellite methods generally do a poor job of representing the site-to-site variations of the annual average $PM_{2.5}$ within the city.

Ulaanbaatar, Mongolia

Ulaanbaatar was another city with several monitor sites, and so generalized additive models (GAMs) were fitted for both the Terra and Aqua data, both generally indicating a poor fit. Figure C.6 shows the evaluation plots for the residuals of the GAM trained on the Aqua data. The response residuals do not follow a normal distribution, which indicates that the statistical fit is fairly poor. Figure C.7 shows that the statistical estimate of ground-level $PM_{2.5}$ increases with the ratio of the satellite AOD to the PBL height for low values of AOD/PBLH, as expected, but the fit is highly unconstrained at larger values, showing what are likely false maxima and minima.

FIGURE C.4

Monthly average PM$_{2.5}$ for each method, Delhi, India

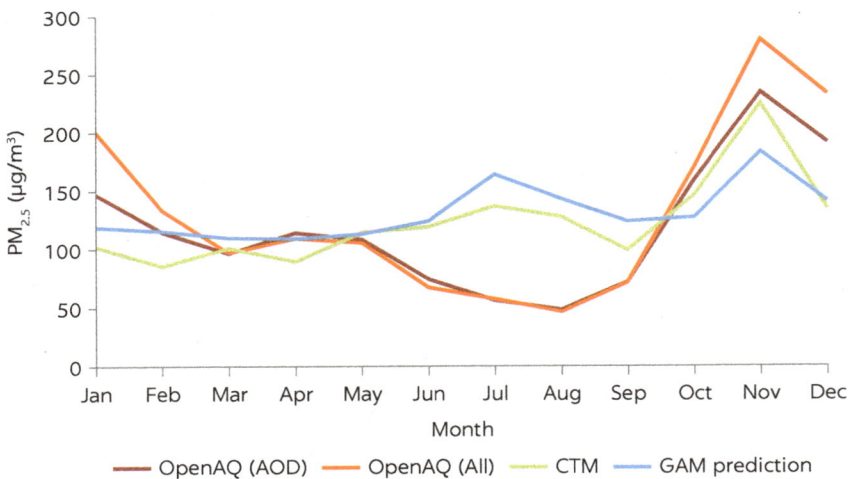

Source: World Bank.
Note: AOD = aerosol optical depth; CTM = chemical transport model; GAM = generalized additive model; OpenAQ = openaq.org; PM$_{2.5}$ = particulate matter with an aerodynamic diameter less than or equal to 2.5 microns; μg/m³ = micrograms per cubic meter.

FIGURE C.5

Annual average PM$_{2.5}$ for each method, by monitor site, Delhi, India

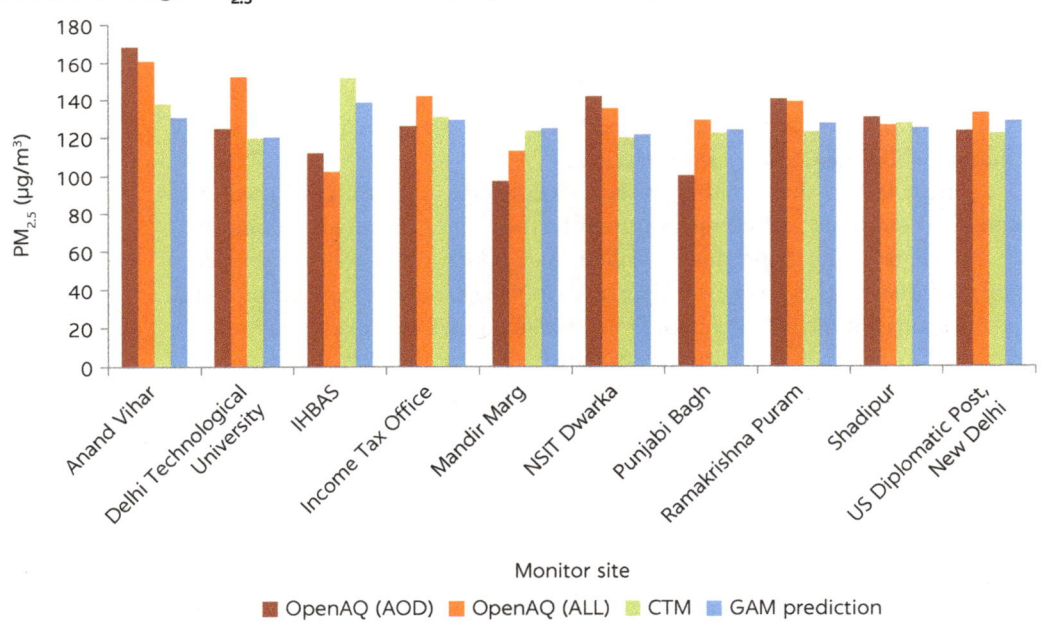

Source: World Bank.
Note: AOD = aerosol optical depth; CTM = chemical transport model; GAM = generalized additive model; NSIT = Netaji Subhas Institute of Technology; OpenAQ = openaq.org; PM$_{2.5}$ = particulate matter with an aerodynamic diameter less than or equal to 2.5 microns; μg/m³ = micrograms per cubic meter.

Table C.4 shows the evaluation statistics from both the statistical and CTM-based methods with respect to the OpenAQ data. It is important to note that no satellite data were available for the winter months in Ulaanbaatar because of persistent snow cover. This means no data were available in

FIGURE C.6

Residual evaluation plots for the generalized additive model trained on the Aqua satellite data, Ulaanbaatar, Mongolia

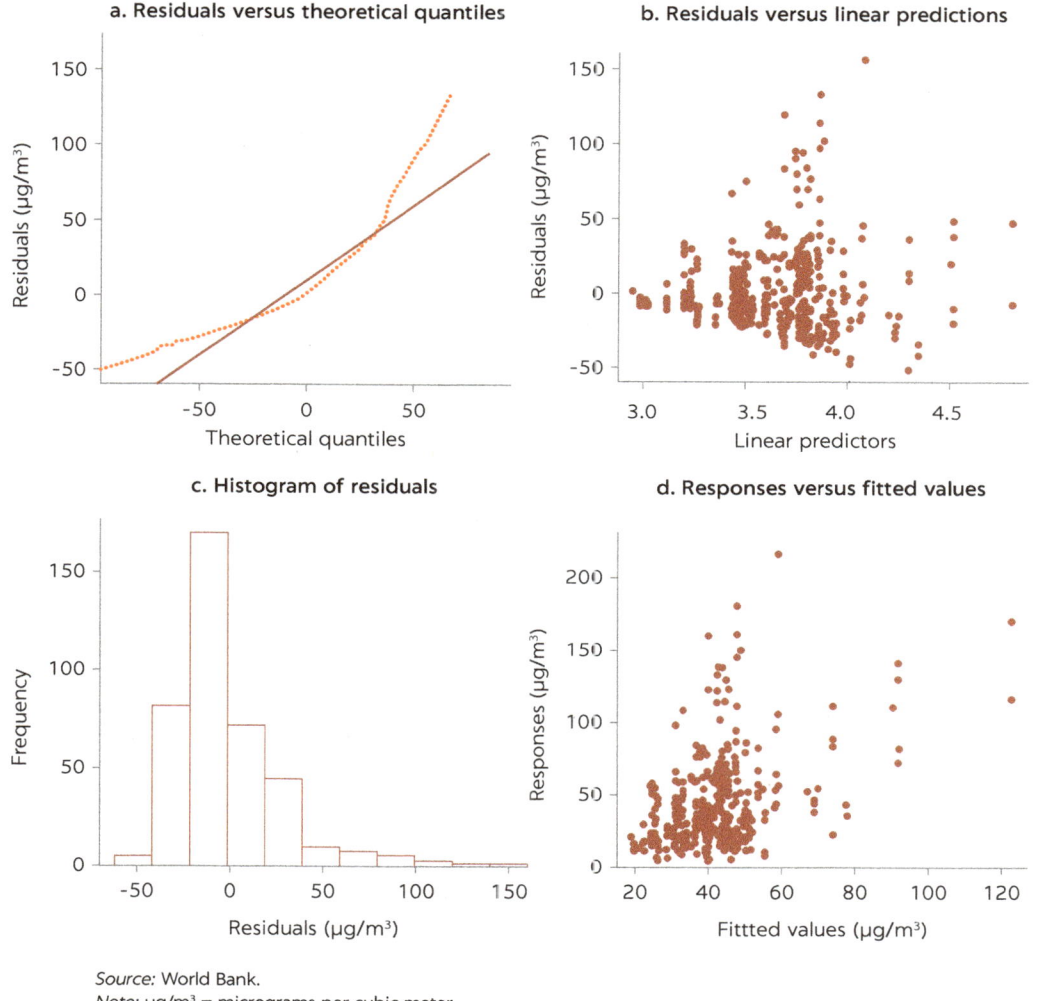

Source: World Bank.
Note: μg/m³ = micrograms per cubic meter.

December–February for the GAM fits or CTM methods to use in the evaluation. The correlation coefficients are fairly low (maximum R of 0.44), and although the mean bias is fairly low for the statistical method as expected, the CTM-based method is consistently biased low, consistent with the results from appendix B.

For Ulaanbaatar, the correlations (R) between the GLM $PM_{2.5}$ concentrations averaged around the satellite overpass time (two-hour window) and the daily average $PM_{2.5}$ concentrations are 66 percent at the Terra overpass time and only 31 percent at the Aqua overpass time. This correlation at the Aqua overpass time is lower than the R for the Aqua GAM in table C.4, which suggests that the statistical approach may be overfitting the data for this case. The high correlation with the Terra overpass time suggests that a better satellite instrument or analysis approach may be possible for this city, but it is unlikely that any new approach or instrument will overcome the loss of data during the snow-covered winter months.

FIGURE C.7

Plot of the dependence of the ground-level PM$_{2.5}$ estimate from the generalized additive model on the ratio of the Aqua aerosol optical depth and the MERRA planetary-boundary-layer height, Ulaanbaatar, Mongolia

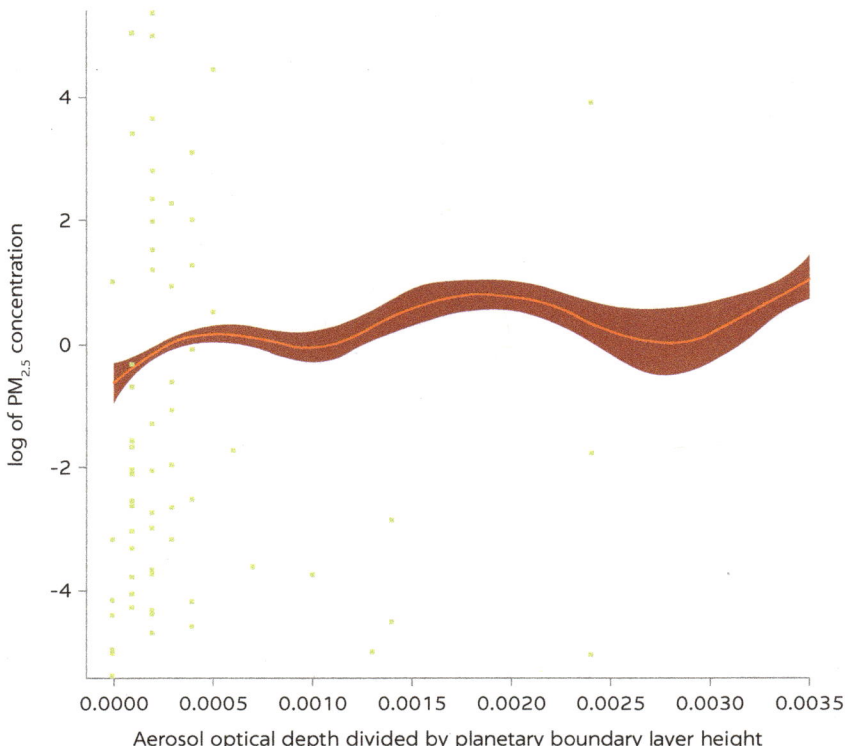

Aerosol optical depth divided by planetary boundary layer height

Source: World Bank.
Note: Green dots = individual observations; red bar = uncertainty of the fit;
MERRA = Modern Era Retrospective-analysis for Research and Applications;
PM$_{2.5}$ = particulate matter with an aerodynamic diameter less than or equal to 2.5 microns.

TABLE C.4 **Evaluation statistics for the generalized-additive-model– and chemical-transport-model–based methods, Ulaanbaatar, Mongolia**

STATISTIC	TERRA		AQUA	
	GAM	CTM	GAM	CTM
Correlation coefficient (*R*)	0.379	0.12	0.442	0.15
Mean bias (micrograms per cubic meter)	0.088	−28.473	0.232	−30.521
Mean normalized bias (%)	52.7	−66.5	53.9	−69.7
Mean normalized gross error (%)	77.2	78.6	78.5	78.1
Root-mean-square error (micrograms per cubic meter)	33.2	48.7	28.9	46.7

Source: World Bank.
Note: CTM = chemical transport model; GAM = generalized additive model.

Figure C.8 shows a scatterplot of the GAM prediction method and the CTM method against the OpenAQ data, respectively, for the US Diplomatic Post monitor site. The CTM-based method is substantially overestimating the concentration. For the statistical method, the bias is lower, but it still shows a very poor fit

FIGURE C.8

Scatterplots of the generalized-additive-model– and chemical-transport-model–based concentrations using the Aqua aerosol-optical-depth data, versus OpenAQ data, US Diplomatic Post in Ulaanbaatar, Mongolia

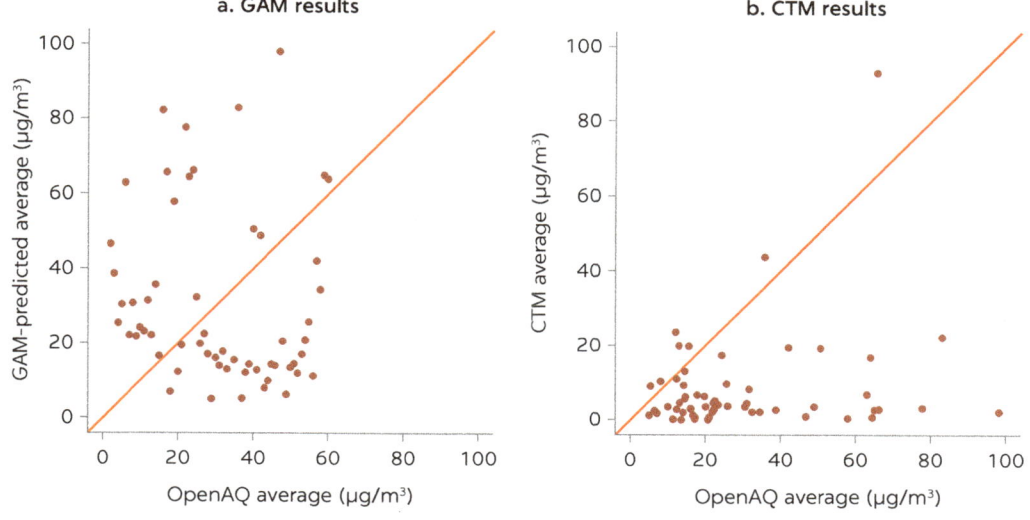

Source: World Bank.
Note: CTM = chemical transport model; GAM = generalized additive model; OpenAQ = openaq.org; µg/m³ = micrograms per cubic meter.

to the data, also reflected in the large mean normalized bias and mean normalized gross error in table C.4.

Figure C.9 shows the citywide monthly averages calculated using each method. This plot shows only the months March through November as stated earlier because of the lack of satellite data during the winter months. Except for July, the CTM-based method greatly underestimates $PM_{2.5}$. The statistical method does a better job of tracking the seasonal cycle of the citywide monthly average, except for November, likely because of low satellite coverage in this month.

Figure C.10 shows the annual average $PM_{2.5}$ for each method by monitoring site. It is reinforced here how the CTM-based method grossly underestimates $PM_{2.5}$ for each site. The statistical method does well except for the sites with the highest values: Bayankhoshuu, Mongolia National Broadcaster, and Tolgoit. This may be due to the influence of local sources of $PM_{2.5}$ at these sites.

Lima, Peru

Lima, Peru, had 10 monitoring sites in the OpenAQ database. However, because of persistent cloud cover over Lima and the exclusion of mixed water and land pixels in the satellite retrieval process, only about 10 valid data points were available over OpenAQ monitors for Aqua and 12 for Terra in the years 2016–17. Figure C.11 shows an aerial view of the Lima area with the OpenAQ monitoring sites in green and the valid satellite AOD data for the two years in blue. Little to no data are available along the coast or over most of Lima, and thus satellite data cannot be used to monitor $PM_{2.5}$ in this city.

Monthly average PM$_{2.5}$ for each method, Ulaanbaatar, Mongolia

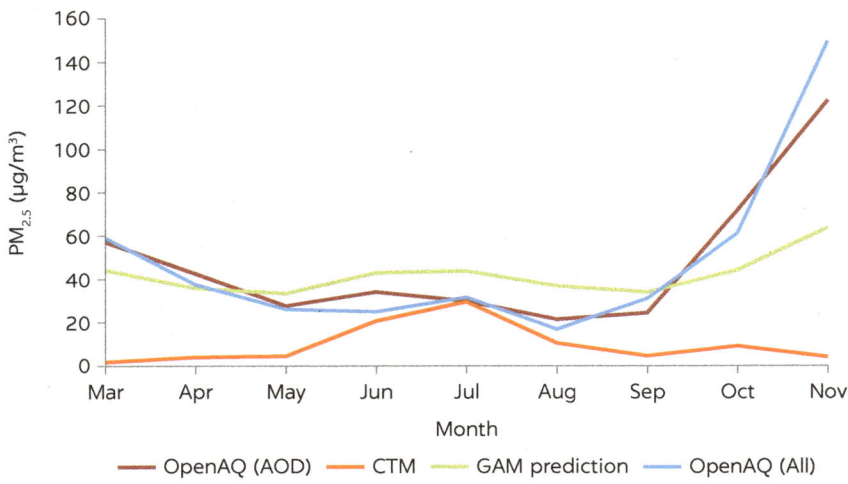

Source: World Bank.
Note: AOD = aerosol optical depth; CTM = chemical transport mode; GAM = generalized additive model; OpenAQ = openaq.org; PM$_{2.5}$ = particulate matter with an aerodynamic diameter less than or equal to 2.5 microns; μg/m³ = micrograms per cubic meter.

Annual average (March–November) PM$_{2.5}$ for each method by monitor site, Ulaanbaatar, Mongolia

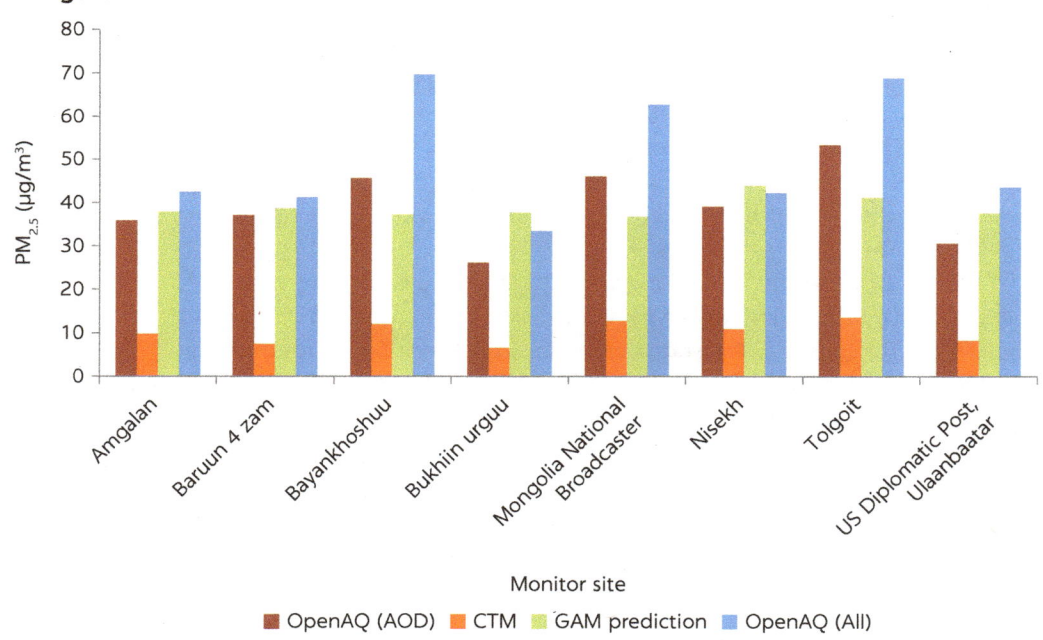

Source: World Bank.
Note: AOD = aerosol optical depth; CTM = chemical transport model; GAM = generalized additive model; OpenAQ = openaq.org; PM = particulate matter. PM$_{2.5}$ = particulate matter with an aerodynamic diameter less than or equal to 2.5 microns; μg/m³ = micrograms per cubic meter.

FIGURE C.11

Valid Terra aerosol-optical-depth retrievals for monitoring sites in the OpenAQ database, Lima, Peru, 2016–17

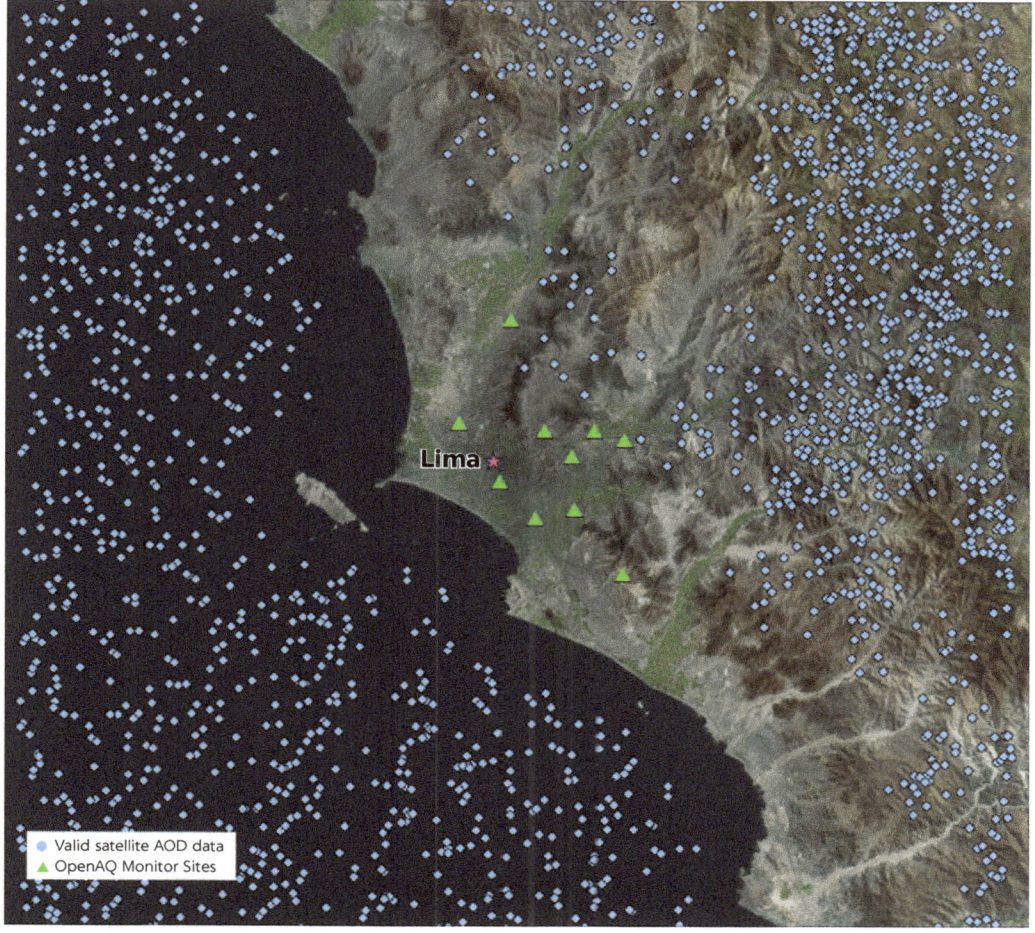

Source: World Bank, produced using Esri ArcGIS.
Note: OpenAQ = openaq.org.

Accra, Ghana

Accra had no monitoring sites in the OpenAQ database, but data for $PM_{2.5}$ concentrations at several sites in 2015 were provided by the Environmental Protection Agency of the Ghana Ministry of Environment, Science, Technology and Innovation and were compared to the CTM-based estimates at these locations.

Table C.5 shows the evaluation statistics for the CTM-based Terra estimates at each site. The correlation coefficients are fairly low (maximum R of 0.4, but for a site where the results are negatively correlated), with significant underestimates of $PM_{2.5}$ at each site and large mean absolute errors. In addition, as shown in the scatterplot in figure C.12, there are very few data points with both valid AOD retrievals and GLM daily averages, and thus attempting a statistical method is unlikely to give much benefit for this city.

TABLE C.5 Evaluation statistics for the chemical-transport-model–based method using Terra data, Accra, Ghana

STATISTIC	WEIJA		TETTEH QUARSHIE INTERCHANGE		MALAM JUNCTION		GRAPHIC	
	TERRA	AQUA	TERRA	AQUA	TERRA	AQUA	TERRA	AQUA
Correlation coefficient (*R*)	0.04	0.40	0.05	0.04	0.12	0.06	0.06	0.04
Mean bias (micrograms per cubic meter)	−44.5	−38.5	−6.7	−16.9	−50.7	−56.4	−25.4	−37.4
Mean normalized bias (%)	−33.3	−22.6	−40.8	−2.9	−46.9	−18.4	−28.5	−15.8
Mean normalized gross error (%)	79.5	55.5	92.3	68.0	51.2	82.5	55.0	71.1
Root-mean-square error (micrograms per cubic meter)	76.1	71.1	63.3	50.0	72.5	84.6	49.3	75.0

Source: World Bank.

FIGURE C.12

Scatterplot of the chemical-transport-model–based PM$_{2.5}$ concentrations using the Terra aerosol-optical-depth data, versus ground-level-monitoring PM$_{2.5}$ data, Malam Junction site in Accra, Ghana

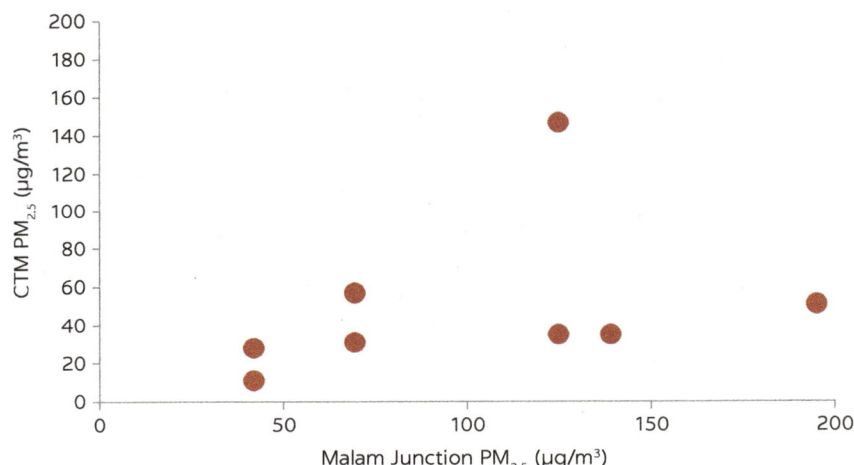

Source: World Bank.
Note: CTM = chemical transport model; PM$_{2.5}$ = particulate matter with an aerodynamic diameter less than or equal to 2.5 microns; μg/m³ = micrograms per cubic meter.

Kathmandu, Nepal

Kathmandu, Nepal, had two OpenAQ monitoring sites, both of which are US Diplomatic Posts, and these sites had data only for 2017. These sites had enough data for the statistical method to be used, but there were only a total of 50 matching satellite-GLM data points, and so monthly and annual averages could not be investigated. Figure C.13 shows the evaluation plots for the residuals of the GAM trained on the Aqua data. The response residuals do follow a normal distribution, which indicates that this statistical fit may be more realistic than those for Delhi and Ulaanbaatar. However, figure C.14 shows that the statistical estimate of ground-level PM$_{2.5}$ does not strongly depend on the ratio of the satellite AOD to the PBL height, with likely unphysical maxima and minima around the relatively low response. This suggests that the statistical fit may be

FIGURE C.13

Residual evaluation plots for the generalized additive model trained on the Aqua satellite data, Kathmandu, Nepal

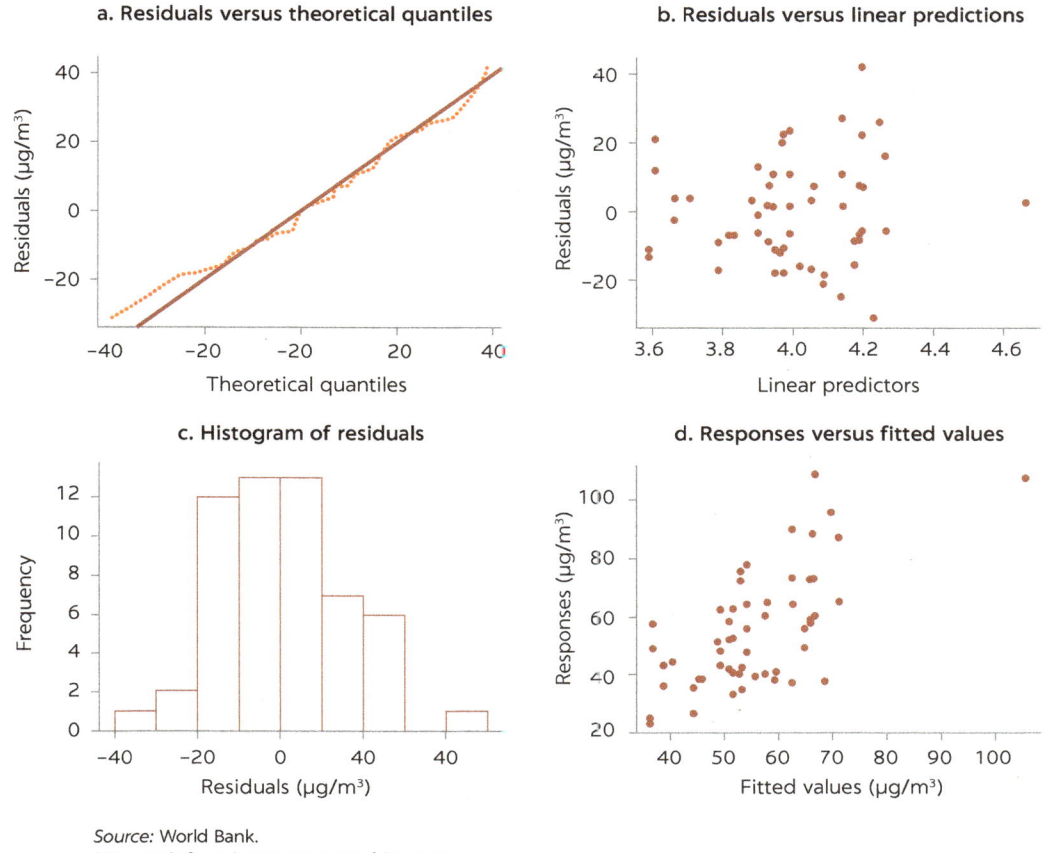

a. Residuals versus theoretical quantiles

b. Residuals versus linear predictions

c. Histogram of residuals

d. Responses versus fitted values

Source: World Bank.
Note: µg/m³ = micrograms per cubic meter.

primarily fitting the errors in the comparison between the AOD/PBLH and the ground-level $PM_{2.5}$ concentrations.

Figure C.15 shows a scatterplot of the statistical (GAM) method and the CTM-based method against the OpenAQ data, respectively, for one of the US Diplomatic Post monitor sites. The statistical method seems to overestimate the monitored values, while the CTM-based method seems to be distributed better but tends to underestimate the values.

Table C.6 shows the evaluation statistics from both the statistical and CTM-based methods with respect to the OpenAQ data. The correlation coefficient for the statistical model trained on the Aqua data is fairly high (R of 0.66), but this is likely an artifact of the relatively low number of matching satellite and GLM data points. This is supported by the fairly low correlation between the GLM data measured at the Aqua overpass time and the GLM daily average $PM_{2.5}$ (R of 32 percent; value of Terra overpass time is 43 percent). The CTM-based method is biased low, similar to the results for Ulaanbaatar, another high-altitude urban site in a mountain valley surrounded by relatively rural land.

FIGURE C.14

Plot of the dependence of the ground-level PM$_{2.5}$ estimate from the generalized additive model on the ratio of the Aqua aerosol optical depth and the MERRA planetary-boundary-layer height, Kathmandu, Nepal

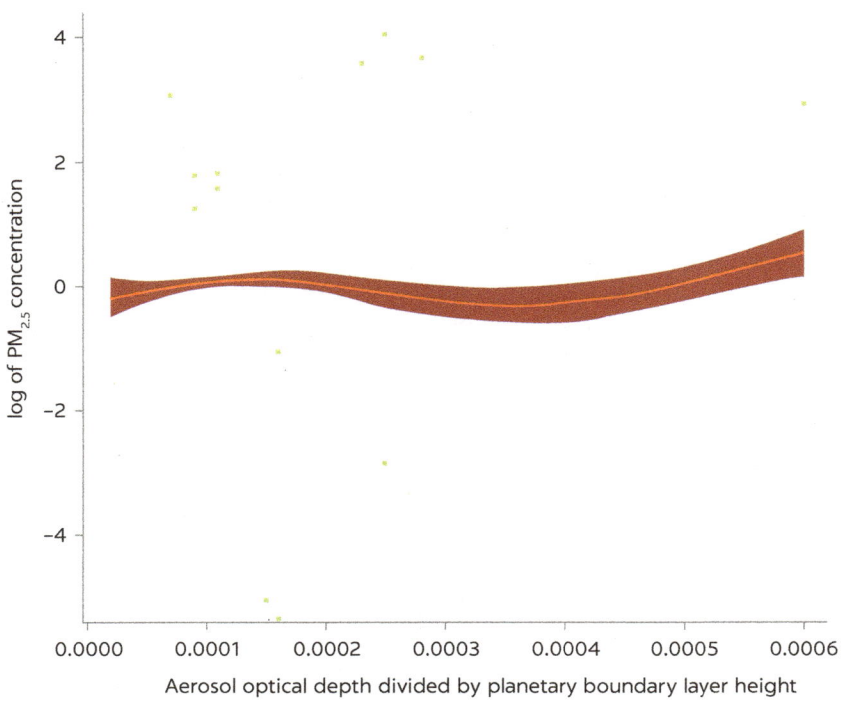

Source: World Bank.
Note: Green dots = individual observations; red bar = uncertainty of the fit;
MERRA = Modern Era Retrospective-analysis for Research and Applications;
PM$_{2.5}$ = particulate matter with an aerodynamic diameter less than or equal to 2.5 microns.

FIGURE C.15

Scatterplots of the generalized-additive-model– and chemical-transport-model–based concentrations using the Aqua aerosol-optical-depth data, versus OpenAQ data, US Diplomatic Post in Kathmandu, Nepal

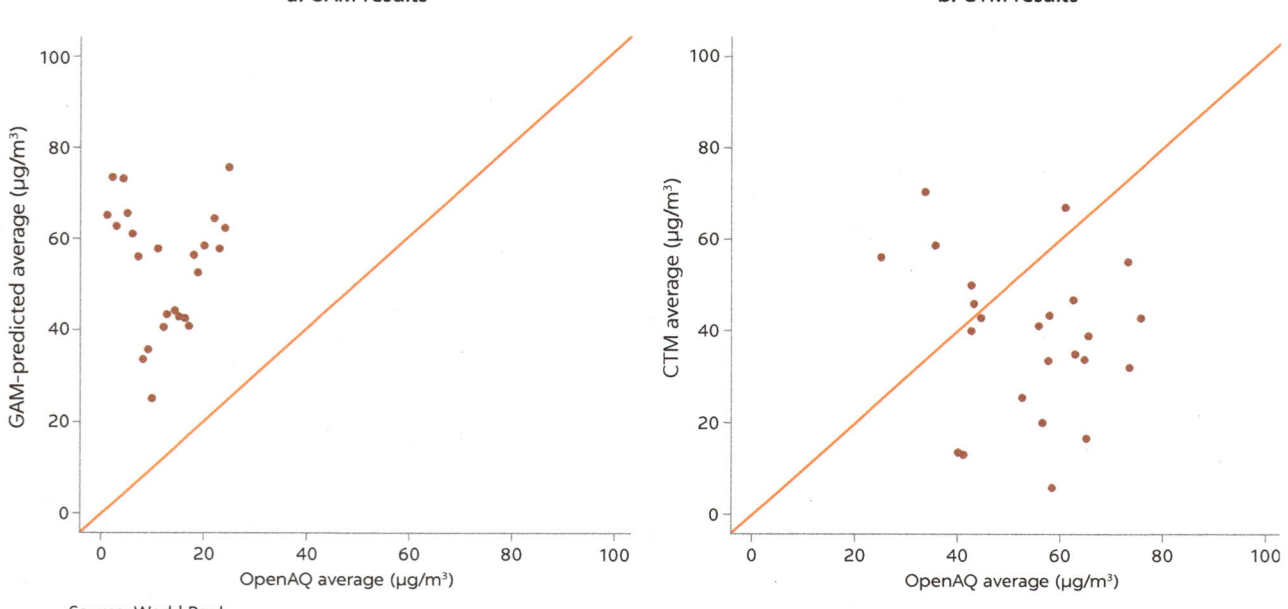

Source: World Bank.
Note: CTM = chemical transport model; GAM = generalized additive model; OpenAQ = openaq.org; μg/m³ = micrograms per cubic meter.

TABLE C.6 Evaluation statistics for the generalized-additive-model– and chemical-transport-model–based methods, Kathmandu, Nepal

	TERRA		AQUA	
STATISTIC	GAM	CTM	GAM	CTM
Correlation coefficient (*R*)	0.479	0.256	0.656	0.130
Mean bias (micrograms per cubic meter)	0.179	−19.768	0.255	−15.801
Mean normalized bias (%)	19.3	−30.1	8.4	−18.5
Mean normalized gross error (%)	39.1	52.1	24.4	47.8
Root-mean-square error (micrograms per cubic meter)	19.1	33.6	14.9	30.1

Source: World Bank.
Note: CTM = chemical transport model; GAM = generalized additive model.

Addis Ababa, Ethiopia

Addis Ababa, Ethiopia, had two sites in the OpenAQ database. Figure C.16 shows the evaluation plots for the residuals of the GAM trained on the Terra data. The response residuals do follow a normal distribution. Figure C.17 shows that the statistical estimate of ground-level $PM_{2.5}$ is linear with the ratio of the satellite AOD to the PBL height, as expected.

FIGURE C.16

Residual evaluation plots for the generalized additive model trained on the Terra satellite data, Addis Ababa, Ethiopia

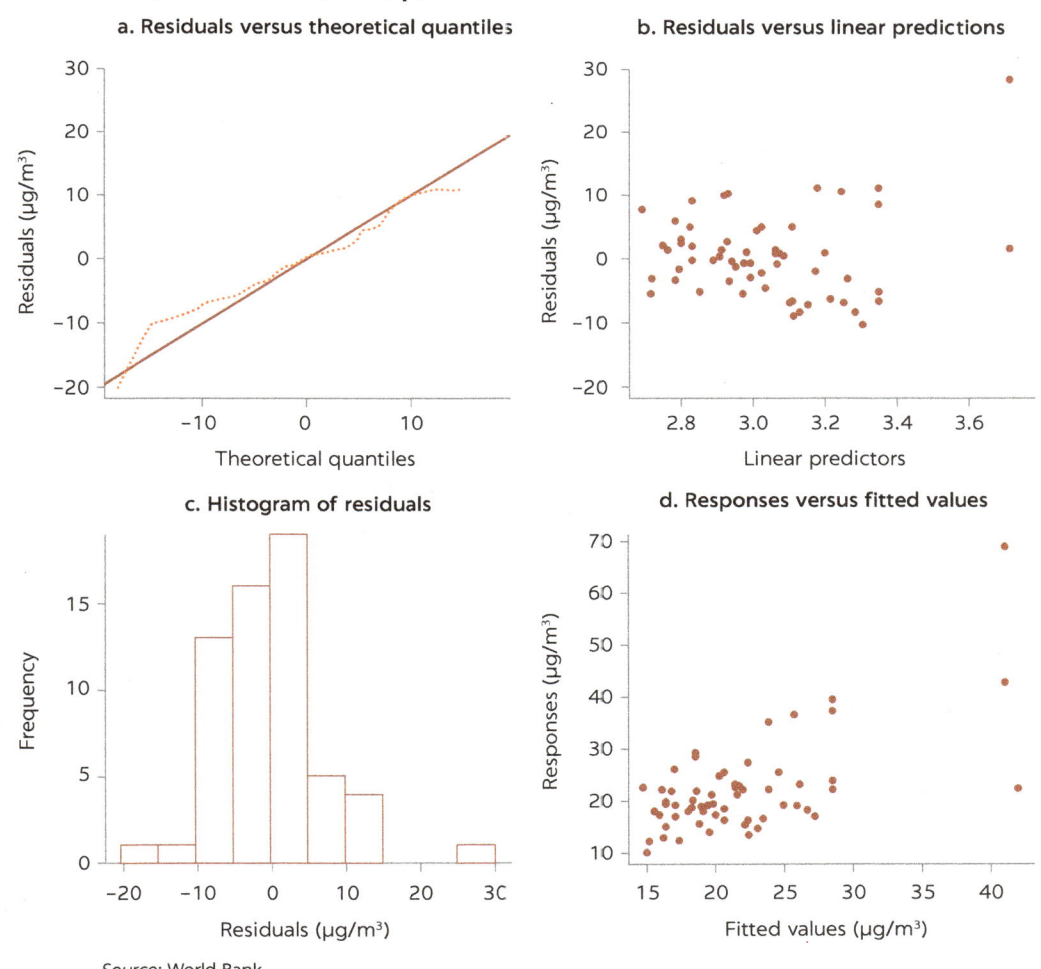

Source: World Bank.
Note: μg/m³ = micrograms per cubic meter.

Figure C.18 shows the scatterplot of the statistical and CTM-based satellite estimates versus the OpenAQ data, and table C.7 shows the statistics from both the statistical and CTM-based methods with respect to the OpenAQ data. The CTM-based estimates tend to be biased high. The GAM trained on Terra data gives a fairly high value for the correlation coefficient, which is slightly higher than the correlation between the GLM data at the Terra overpass time and the GLM daily average $PM_{2.5}$ (59 percent; 81 percent for Aqua overpass time), suggesting that the fit may be due to the low number of points with a valid AOD estimate.

FIGURE C.17

Plot of the dependence of the ground-level $PM_{2.5}$ estimate from the generalized additive model on the ratio of the Terra aerosol optical depth and the MERRA planetary-boundary-layer height, Addis Ababa, Ethiopia

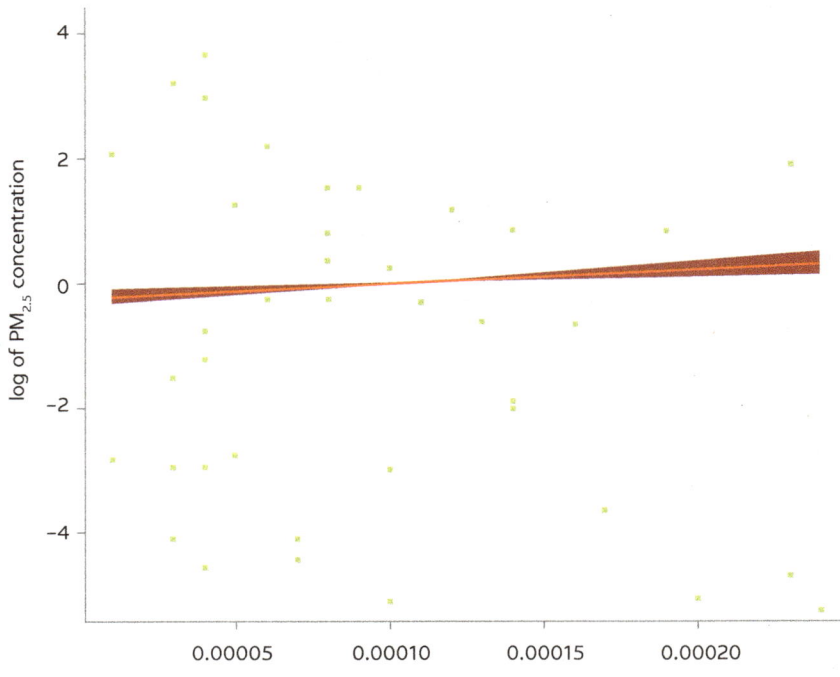

Source: World Bank.
Note: Green dots = individual observations; red bar = uncertainty of the fit;
MERRA = Modern Era Retrospective-analysis for Research and Applications;
$PM_{2.5}$ = particulate matter with an aerodynamic diameter less than or equal to 2.5 microns.

TABLE C.7 **Evaluation statistics for the generalized-additive-model– and chemical-transport-model–based methods, Addis Ababa, Ethiopia**

	TERRA		AQUA	
STATISTIC	GAM	CTM	GAM	CTM
Correlation coefficient (R)	0.63	0.399	0.441	0.054
Mean bias (micrograms per cubic meter)	−0.051	6.463	0.0003	15.786
Mean normalized bias (%)	6.9	35.5	7.4	92.2
Mean normalized gross error (%)	23.5	64.8	21.3	104.5
Root-mean-square error (micrograms per cubic meter)	7.0	17.5	6.2	28.9

Source: World Bank.
Note: CTM = chemical transport model; GAM = generalized additive model.

FIGURE C.18

Scatterplots of the generalized-additive-model– and chemical-transport-model–based concentrations, using the Aqua aerosol-optical-depth data, versus OpenAQ Data, US Diplomatic Post in Addis Ababa, Ethiopia

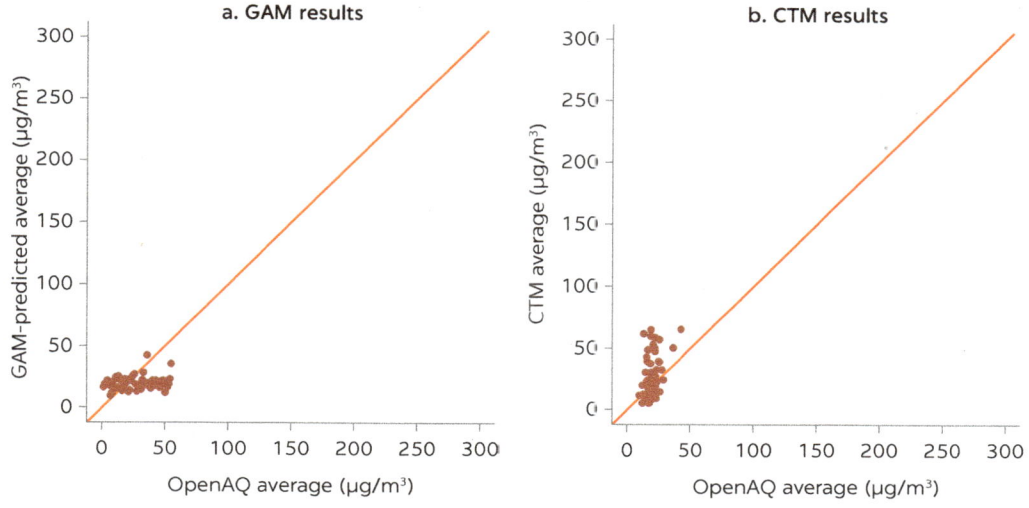

Source: World Bank.
Note: CTM = chemical transport model; GAM = generalized additive model; OpenAQ = openaq.org; µg/m³ = micrograms per cubic meter.

Dakar, Senegal

Figure C.19 shows a scatterplot of the Dakar $PM_{2.5}$ data for 2011 versus the CTM-based satellite estimate using the Terra data. Since the Dakar GLM $PM_{2.5}$ data are the average of four sites, the CTM-based estimates were calculated for each site and then averaged before comparing them with the GLM data. Two valid AOD values (and thus valid CTM-based $PM_{2.5}$ estimates) were required to consider the satellite average as a valid data point.

FIGURE C.19

Scatterplot of the chemical-transport-model–based $PM_{2.5}$ concentrations using the Terra aerosol-optical-depth data versus citywide average $PM_{2.5}$ data, Dakar, Senegal

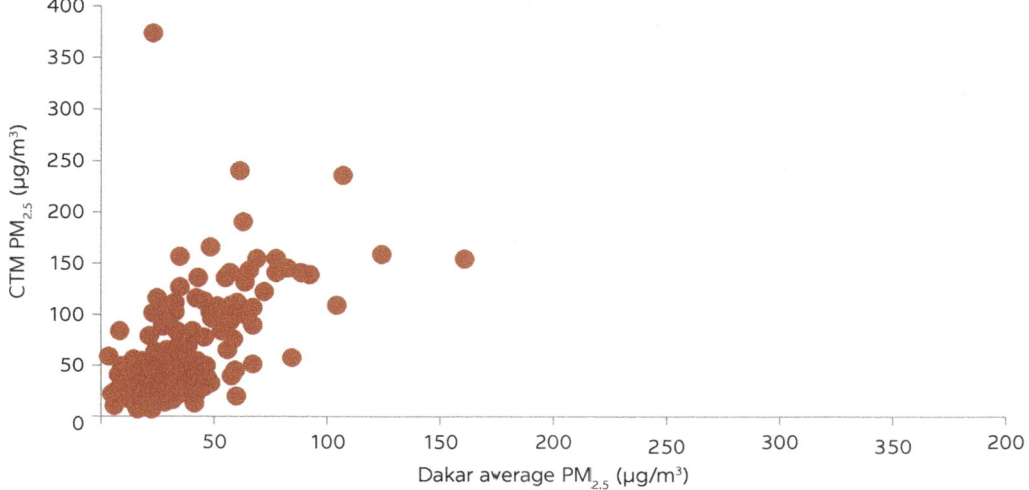

Source: World Bank.
Note: CTM = chemical transport model; $PM_{2.5}$ = particulate matter with an aerodynamic diameter less than or equal to 2.5 microns; µg/m³ = micrograms per cubic meter.

TABLE C.8 Evaluation statistics for the chemical-transport-model–based method, Dakar, Senegal

STATISTIC	TERRA	AQUA
Correlation coefficient (R)	0.35	0.52
Mean bias (micrograms per cubic meter)	24.3	23.0
Mean normalized bias (%)	86.0	67.5
Mean normalized gross error (%)	101.6	85.3
Root-mean-square error (micrograms per cubic meter)	45.3	41.4

Source: World Bank.

Table C.8 shows the statistics for the comparison. The correlation coefficient is fairly high (R of 52 percent), but the CTM-based approach tends to overestimate the $PM_{2.5}$ concentrations and has very large values for mean normalized bias and mean normalized gross error (86 percent and 101 percent), indicating fairly poor performance in this city.

Kampala, Uganda

Kampala, Uganda, has one OpenAQ site, but there were very few valid daily averages in the OpenAQ data, which combined with the relatively low coverage of the AOD retrievals left few matching points for evaluation. There were only 14 matching points for the Terra data, and only 22 for Aqua, so only the Aqua data are discussed further here. The low number of points in the statistical fit for the Aqua data do not give us confidence in the results. The GAM evaluation plots in figure C.20 look reasonable, as does the linear dependence of the ground-level $PM_{2.5}$ estimate on AOD/PBLH in figure C.21, but the low number of valid AOD points suggests that this may be coincidental.

Figure C.22 shows the scatterplot of the statistical and CTM-based satellite estimates versus the OpenAQ data, and table C.9 shows the statistics from both the statistical and CTM-based methods with respect to the OpenAQ data. The GAM seems to perform well, but this is likely due to the low number of points with a valid AOD. The correlation coefficient for the GAM is larger than the correlations between the GLM $PM_{2.5}$ data at the Aqua overpass time and the daily average $PM_{2.5}$ (45 percent). In contrast, the CTM-based estimates appear to be biased low even for the small number of points available, and the correlation coefficient is very poor (14 percent).

FIGURE C.20

Residual evaluation plots for the generalized additive model trained on the Aqua satellite data, Kampala, Uganda

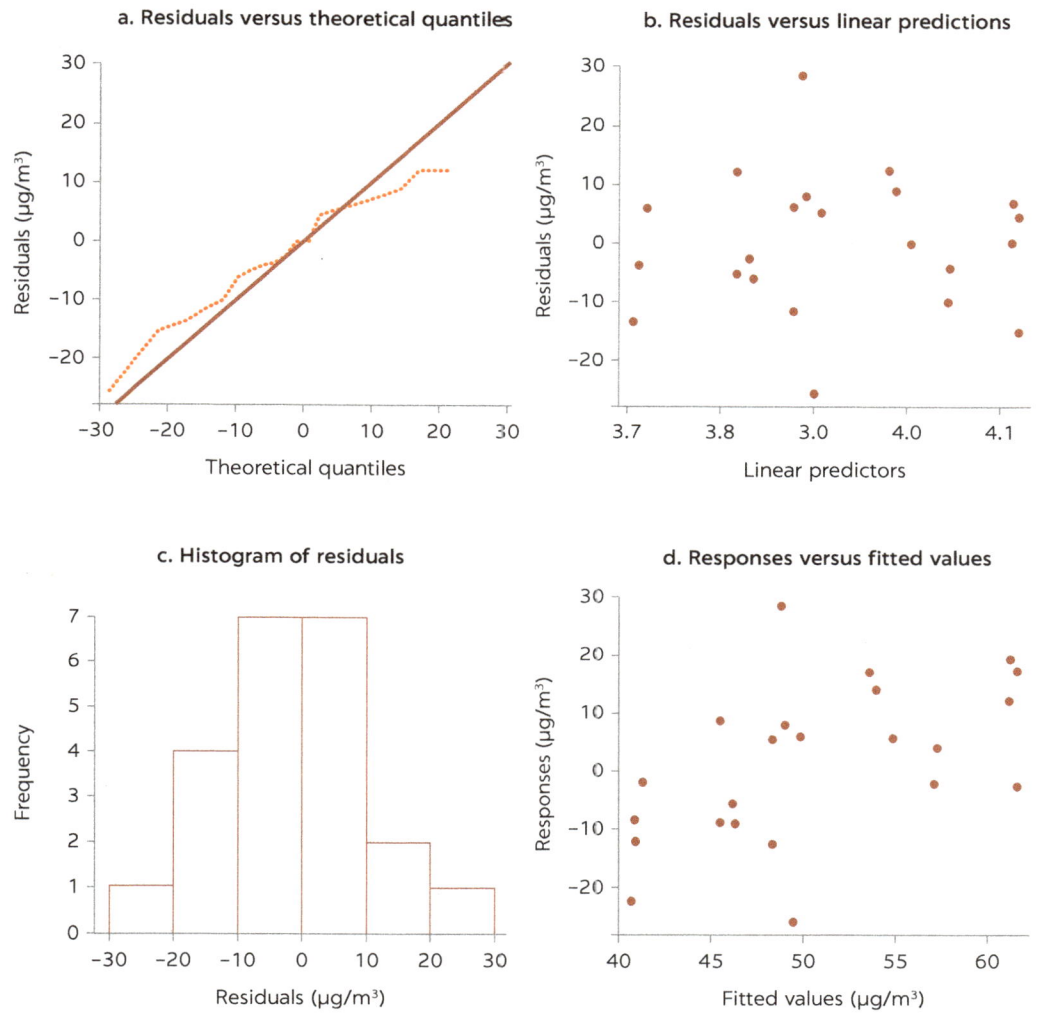

Source: World Bank.
Note: µg/m³ = micrograms per cubic meter.

FIGURE C.21

FIGURE C.21

Plot of the dependence of the ground-level PM$_{2.5}$ estimate from the generalized additive model on the ratio of the Aqua aerosol optical depth and the MERRA planetary-boundary-layer height, Kampala, Uganda

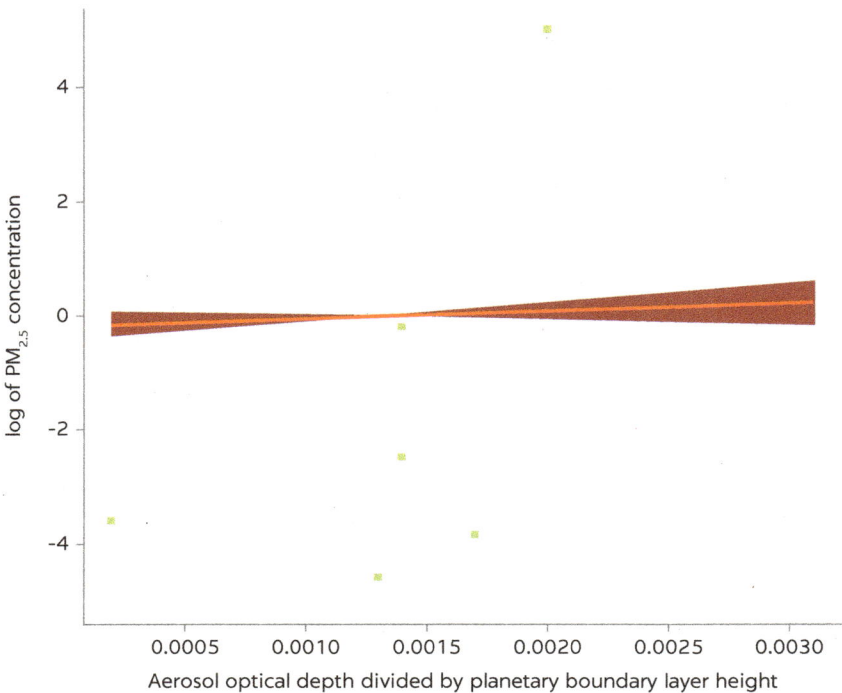

Source: World Bank.
Note: Green dots = individual observations; red bar = uncertainty of the fit; MERRA = Modern Era Retrospective-analysis for Research and Applications; PM$_{2.5}$ = particulate matter with an aerodynamic diameter less than or equal to 2.5 microns.

FIGURE C.22

Scatterplots of the generalized-additive-model– and chemical-transport-model–based PM$_{2.5}$ concentrations, using the Aqua aerosol-optical-depth data, versus OpenAQ data, US Diplomatic Post, Kampala, Uganda

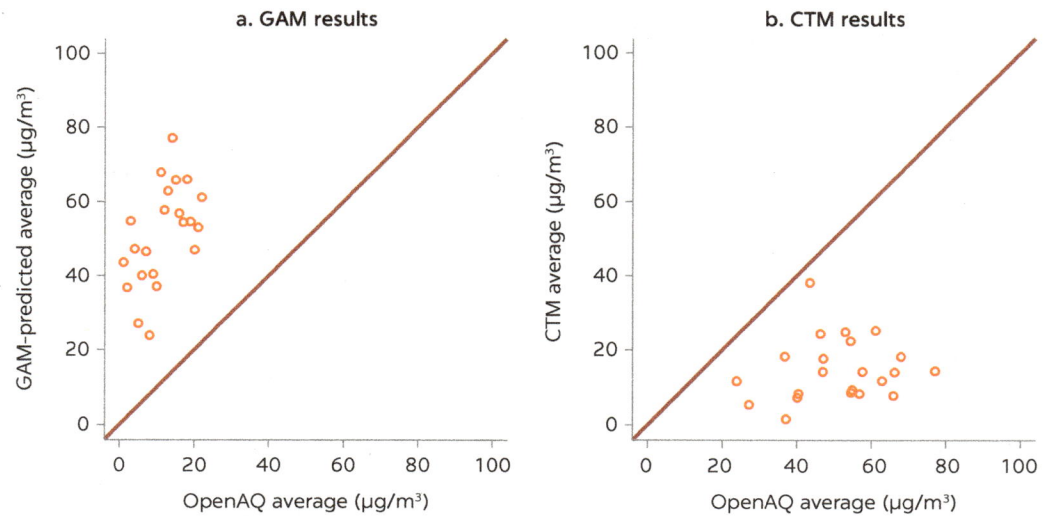

Source: World Bank.
Note: CTM = chemical transport model; GAM = generalized additive model; OpenAQ = openaq.org; µg/m³ = micrograms per cubic meter.

TABLE C.9 Evaluation statistics for the generalized-additive-model- and chemical-transport-model-based methods, Kampala, Uganda

STATISTIC	AQUA	
	GAM	CTM
Correlation coefficient (*R*)	0.514	0.139
Mean bias (micrograms per cubic meter)	0.018	−36.16
Mean normalized bias (%)	6.5	−69.5
Mean normalized gross error (%)	20.5	69.5
Root-mean-square error (micrograms per cubic meter)	11.3	39

Source: World Bank.
Note: CTM = chemical transport model; GAM = generalized additive model.

Hanoi, Vietnam

Hanoi, Vietnam, had one site in the OpenAQ data; it was possible to complete both Terra and Aqua GAM fits. The GAM evaluation plots in figure C.23 look reasonable, as does the near linear dependence of the ground-level PM$_{2.5}$ estimate

FIGURE C.23

Residual evaluation plots for the generalized additive model trained on the Terra satellite data, Hanoi, Vietnam

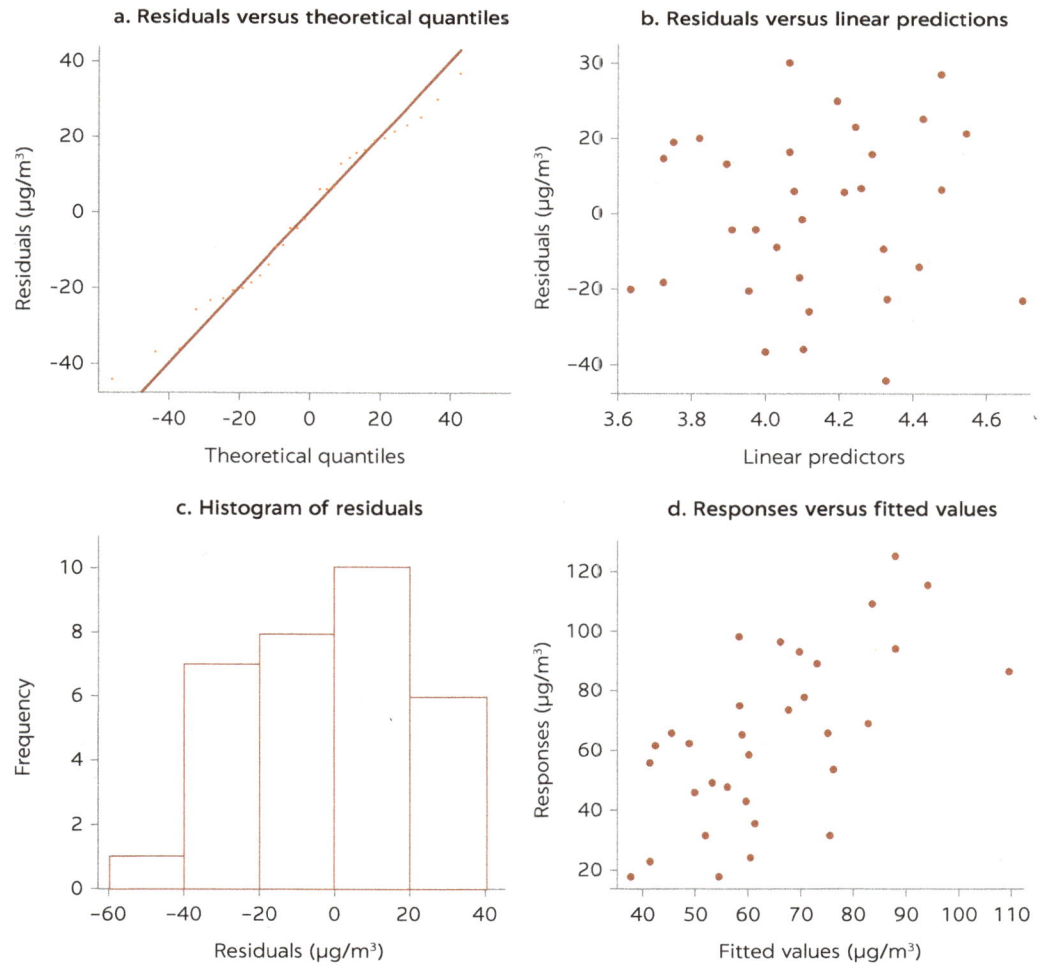

Source: World Bank.
Note: μg/m³ = micrograms per cubic meter.

on AOD/PBLH in figure C.24, but the low number of valid matching GLM and AOD points (less than 33 for all cases) suggests that this may be coincidental.

Figure C.25 shows the scatterplot of the statistical and CTM-based satellite estimates versus the OpenAQ data, and table C.10 shows the statistics from both the statistical and CTM-based methods with respect to the OpenAQ data. The GAM seems to perform well, but this may be due to the low number of points with a valid AOD. However, the correlation coefficient for the GAM is smaller than the correlations between the GLM $PM_{2.5}$ data at the Terra overpass time and the daily average $PM_{2.5}$ (R of 73 percent; 71 percent for Aqua). The CTM-based estimates are also relatively unbiased, although the correlation coefficient is low (30 percent).

FIGURE C.24

Plot of the dependence of the ground-level PM$_{2.5}$ estimate from the generalized additive model on the ratio of the Terra aerosol optical depth and the MERRA planetary-boundary-layer height, Hanoi, Vietnam

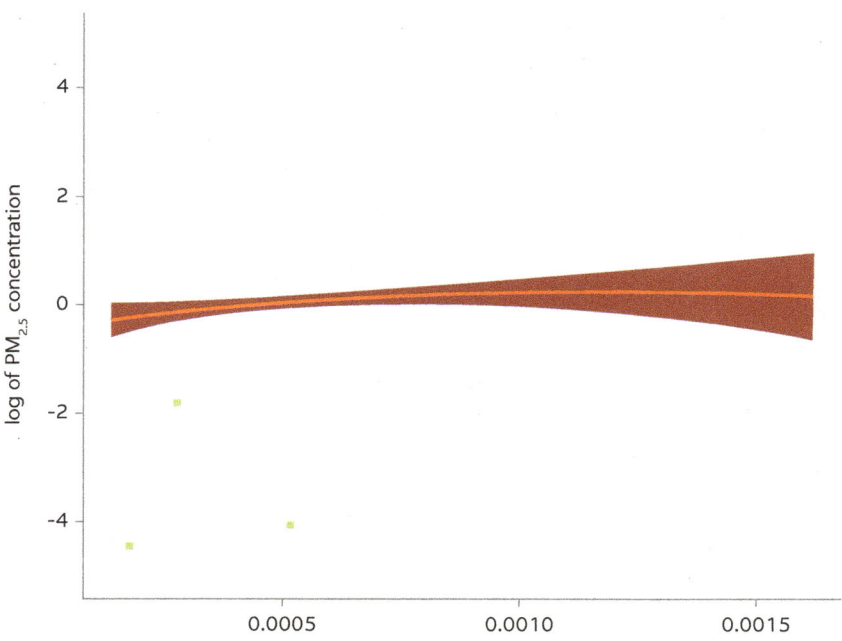

Source: World Bank.
Note: Green dots = individual observations; red bar = uncertainty of the fit;
MERRA = Modern Era Retrospective-analysis for Research and Applications;
$PM_{2.5}$ = particulate matter with an aerodynamic diameter less than or equal to 2.5 microns.

FIGURE C.25

Scatterplots of the generalized-additive-model– and chemical-transport-model–based concentrations using Terra aerosol-optical-depth data, versus OpenAQ data, US Diplomatic Post in Hanoi, Vietnam

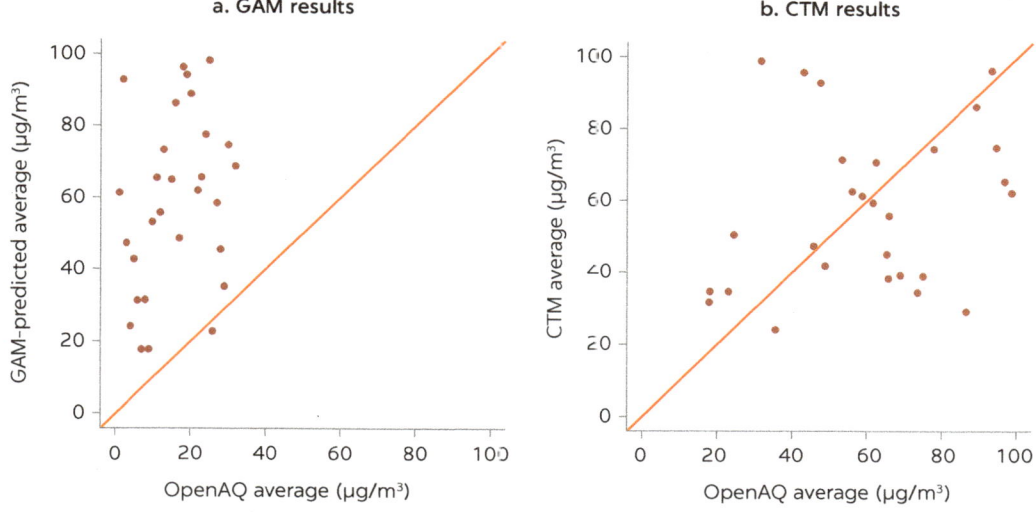

Source: World Bank.
Note: CTM = chemical transport model; GAM = generalized additive model; OpenAQ = openaq.org; µg/m³ = micrograms per cubic meter.

TABLE C.10 Evaluation statistics for the generalized-additive-model– and chemical-transport-model–based methods, Hanoi, Vietnam

STATISTIC	TERRA		AQUA	
	GAM	CTM	GAM	CTM
Correlation coefficient (R)	0.634	0.302	0.626	0.259
Mean bias (micrograms per cubic meter)	0.347	0.175	0.76	6.367
Mean normalized bias (%)	20.9	21.8	22.2	39.6
Mean normalized gross error (%)	42.5	53.4	39.1	67.8
Root-mean-square error (micrograms per cubic meter)	22.0	34.3	21.0	31.0

Source: World Bank.
Note: CTM = chemical transport model; GAM = generalized additive model.

OPENAQ QUALITY CHECK TOOL

OpenAQ worked with the organization Development Seed to create a command line tool[1] that can aid in quality checking the OpenAQ data. This tool, called openaq-quality-checks (QA [quality assurance] Tool), can flag each data point based on metrics given by the user, for example, missing values, negative values, or values greater than some user-provided threshold. This tool was applied to the Delhi data to see the change in results when additional quality measures are applied. The tool was used to remove values noted as missing (–999), values less than 1, values over a provided threshold, and data points where the same value was repeated for multiple consecutive hours at the same site. The data for this city is subhourly. At 15-minute intervals, it makes sense that the concentration values may be the same. Therefore, the repeat option for this tool is not valid for this data set. For this exercise the tool was run to remove only values less than 1 (including missing values) and values over a calculated threshold. This threshold was derived by finding the standard deviation of the log of hourly values for all sites for the two years and the mean of the log for all hourly values. The concentration threshold was calculated as three times this standard deviation plus the mean of the log. The QA Tool then removes any values over this threshold. For Delhi the threshold is 1,223 micrograms per cubic meter. This does not prove that points above this threshold actually represent poor data but instead merely tests the impact of applying this threshold to the analysis. Furthermore, the initial processing already removed negative and zero values, and so the main impact of the QA Tool is expected to be from applying this upper threshold.

Figure C.26 shows the difference in the monthly averages of the base processing done to all sites (here for Delhi) and the QA Tool equivalent. It is apparent the US Diplomatic Post had unusually high values during a few months that the QA Tool removed. Table C.11 shows the 2016–17 average PM$_{2.5}$ concentrations for each OpenAQ site in Delhi. Consistent with figure C.26, the only significant changes are for the US Diplomatic Post, where large values were filtered.

FIGURE C.26

Difference in base processing monthly average versus monthly averages after the quality assurance tool is applied

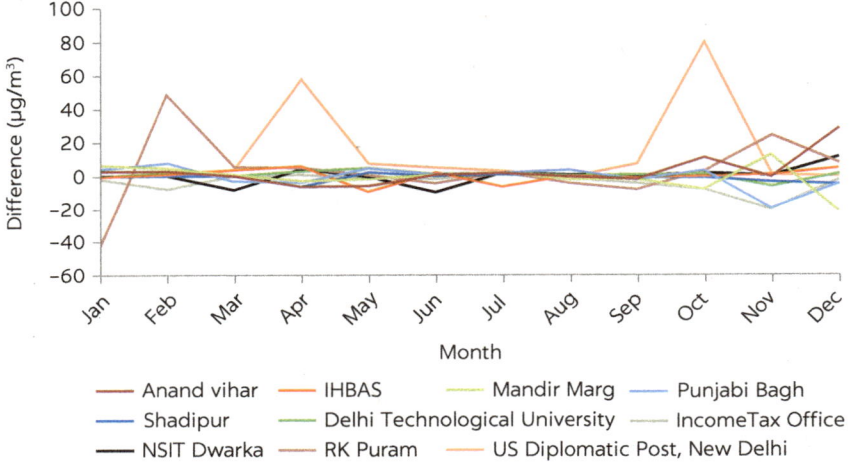

Source: World Bank.
Note: IHBAS = Institute of Human Behavior and Allied Sciences; NSIT = Netaji Subhas Institute of Technology; RK Puram = Ramakrishna Puram; μg/m³ = micrograms per cubic meter.

TABLE C.11 Average PM$_{2.5}$ concentrations for each monitor site in Delhi, India, 2016–17

SITE	QUALITY ASSURANCE TOOL	BASE PROCESSING
Anand Vihar	159.2	160.7
Delhi Technological University	156.3	152.2
Institute of Human Behavior and Allied Sciences	94.7	102.2
Income Tax Office	144.4	141.4
Mandir Marg	115.6	112.8
Netaji Subhas Institute of Technology Dwarka	132.0	134.6
Punjabi Bagh	129.8	128.4
Ramakrishna Puram	136.6	138.3
Shadipur	127.2	125.7
US Diplomatic Post: New Delhi	116.3	132.0

Source: World Bank.
Note: PM$_{2.5}$ = particulate matter with an aerodynamic diameter less than or equal to 2.5 microns.

However, the investigation into the quality assurance and control (QA/QC) processes for the US Diplomatic Post data showed that they are following US EPA procedures for QA/QC and instrument maintenance, so it would seem unlikely that these points truly represent poor quality data.

Using these data from the QA Tool, the same analysis scripts as before were run to merge the satellite data and train a GAM. Table C.12 shows the fit statistics of the GAM using the QA Tool data versus the data used in the base processing. Both runs with the QA Tool (Terra and Aqua) show a better fit than the base processing runs; however, an *R* value of 0.3 is still a poor fit. Figure C.27

TABLE C.12 **Generalized-additive-model output statistics of quality assurance tool versus base processing, Delhi, India**

	BASE PROCESSING		QUALITY ASSURANCE TOOL	
STATISTIC	TERRA	AQUA	TERRA	AQUA
Correlation coefficient (*R*)	0.204	0.102	0.3	0.196
Deviance explained	20.8%	10.7%	30.5%	20.0%

Source: World Bank.

FIGURE C.27

Generalized-additive-model residual evaluation plots from Terra quality assurance tool

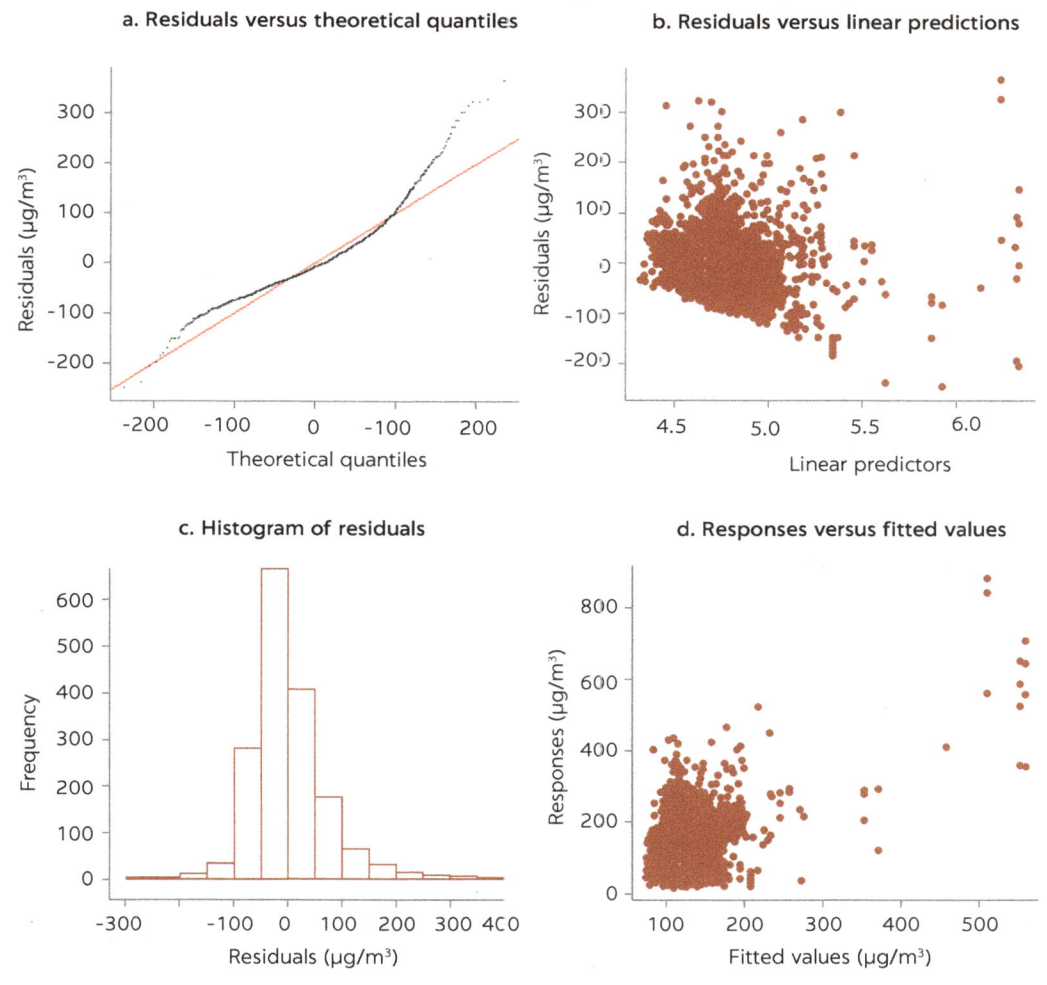

Source: World Bank.
Note: μg/m³ = micrograms per cubic meter.

shows the GAM residual evaluation plots from the Terra QA Tool run, which can be compared to figure C.1 from the base processing run. The residuals are much closer to a normal distribution after the QA filter is applied.

Figure C.28 shows a scatterplot of the ground-level PM$_{2.5}$ estimate from the CTM-based method and the statistical method, respectively, against the filtered OpenAQ data for the US Diplomatic Post monitor site (compare with figure C.3). Table C.13 shows the evaluation statistics from both the statistical and CTM-based methods with respect to the filtered OpenAQ data (compare with table C.3). The correlation coefficients have increased somewhat (maximum R from 0.46 to 0.55) but that the overall bias and error statistics have not significantly changed.

FIGURE C.28

Scatterplots of the statistical and chemical-transport-model–based concentrations using the Aqua aerosol-optical-depth data versus the QA filtered OpenAQ data, US Diplomatic Post in Ulaanbaatar, Mongolia

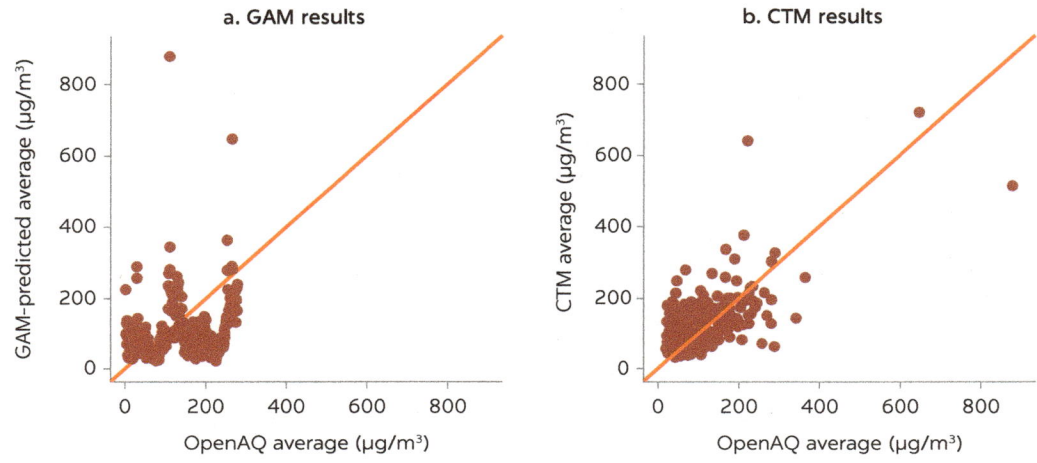

Source: World Bank.
Note: CTM = chemical transport model; GAM = generalized additive model; OpenAQ = openaq.org; QA = quality assurance; µg/m³ = micrograms per cubic meter.

TABLE C.13 **Evaluation statistics (using the filtered OpenAQ data) for the statistical and chemical-transport-model–based methods, Delhi, India**

STATISTIC	TERRA		AQUA	
	GAM	CTM	GAM	CTM
Correlation coefficient (*R*)	0.552	0.537	0.447	0.354
Mean bias (micrograms per cubic meter)	−0.349	2.642	0.078	−27.483
Mean normalized bias (%)	34.2	28.0	26.1	−6.1
Mean normalized gross error (%)	57.5	60.2	47.7	48.1
Root-mean-square error (micrograms per cubic meter)	68.2	78.8	56.5	71.8

Source: World Bank.
Note: CTM = chemical transport model; GAM = generalized additive model; OpenAQ = openaq.org.

CO-KRIGING AND KRIGING WITH SATELLITE AND OPENAQ DATA

The goal of this test was to determine if satellite data, although not a replacement for GLM data, could be combined with GLM data in a way that reduces the number of GLM sites required to quantify exposure in an urban population. To test this, ordinary kriging was done for each day using the monitor station daily averages, excluding the US Diplomatic Post. The concentration at the US Diplomatic Post for each day was then predicted with ordinary kriging. This tested how well the other GLM sites alone could characterize the PM$_{2.5}$ concentrations at an unmeasured site. The CTM-based satellite PM$_{2.5}$ estimates were then combined with the GLM data (minus the US Diplomatic Post) via co-kriging and again estimated the concentrations at the US Diplomatic Post. The results for Delhi and Ulaanbaatar are shown in tables C.14 and C.15, respectively.

The Delhi results suggest that adding in the satellite data would not reduce the number of monitoring sites there, because the root mean square is actually higher when the CTM-based satellite data are added. The Ulaanbaatar result suggests that the Diplomatic Post could be eliminated by adding satellite data to the rest of the GLM network in this city, but again only for the eight months a year that satellites actually produce data.

TABLE C.14 **Kriging and co-kriging statistics, Delhi, India**

STATISTIC	CO-KRIGING	KRIGING
Mean bias (micrograms per cubic meter)	−4.330	−5.935
Root-mean-square error (micrograms per cubic meter)	148.527	134.374

Source: World Bank.

TABLE C.15 **Kriging and co-kriging statistics, Ulaanbaatar, Mongolia**

STATISTIC	CO-KRIGING	KRIGING
Mean bias (micrograms per cubic meter)	−1.670	−12.400
Root-mean-square error (micrograms per cubic meter)	27.558	55.204

Source: World Bank.

NOTE

1. https://github.com/openaq/openaq-quality-checks.

Quality Considerations for Ground-Level-Monitoring Data in Low- and Middle-Income Countries

This appendix provides further details on the uncertainties in ground-level-monitoring (GLM) data in low- and middle-income countries (LMICs). There are two points of focus here: (1) a focus on the uncertainties associated with two commonly used measurement techniques; that is, the beta attenuation mass monitors (first section) and the tapered element oscillating microbalance (second section), and (2) a focus on evaluation of the consistency between the GLM data collected by the US Diplomatic Posts and the GLM data collected by local authorities in Delhi, India (third section) and Lima, Peru (final section).

OPERATING METHODOLOGIES AND UNCERTAINTIES ASSOCIATED WITH THE BAMM TECHNIQUE

Beta attenuation mass monitors (BAMMs), including the Met One model BAM-1020 and the Thermo Fisher Scientific BAMM model 5014i, are the most common instruments for monitoring particulate matter with an aerodynamic diameter less than or equal to 2.5 microns ($PM_{2.5}$) across the globe. US Diplomatic Posts deploy these instruments for $PM_{2.5}$ measurements.

A $PM_{2.5}$ measurement system typically includes a sample inlet, a PM cyclone, a sample conditioning unit (such as a heater to ensure modest relative humidity inside the instrument), the $PM_{2.5}$ instrument, and a pump to draw a sample through the system.

The beta attenuation monitor (BAM) technique samples ambient PM onto a filter and measures the attenuation of beta rays (high-energy electrons) through the PM-loaded filter in time. The beta rays emitted from a radioactive source (^{14}C) are attenuated (absorbed) by PM collected on a filter:

$$I = I_o e^{-\mu x} \text{ and } x = \frac{1}{\mu}\ln\left[\frac{I_o}{I}\right], \tag{D.1}$$

where I is the measured beta ray intensity (counts per unit of time) of the attenuated beta ray (particle-laden filter tape), I_o is the measured beta ray intensity of the nonattenuated beta ray (clean filter tape), μ is the beta ray attenuation cross section of the PM (square meters per kilogram), and x is the mass density of the PM (kilograms per square meter) on the filter. Key to the success of the beta attenuation monitor is, in part, that μ, the absorption cross section, varies little among commonly sampled PM such as C, Fe_2O_3, NH_4NO_3, $NH_4(SO_4)_2$, or SiO_2. This permits the device to be calibrated during the manufacturing process and permits the user to measure PM concentrations without having to know the chemical composition of the sampled PM in advance.

$PM_{2.5}$ mass concentrations are calculated from the measured attenuation. The ambient PM concentration of particulate matter (kilograms per cubic meter) is

$$C(kg/m^3) = \frac{x(kg/m^2)A(m^2)}{Q(m^3/s) \times t(s)} \quad \text{and} \quad C(\mu g/m^3) = \frac{D}{\mu} \ln\left[\frac{I_o}{I}\right], \qquad \text{(D.2)}$$

where C is the ambient PM concentration (kilograms per cubic meter), A is the cross-sectional area of the tape spot over which PM is being deposited (m^2), Q is the rate at which ambient air is being sampled (cubic meters per second), t is the sampling time (s), and D is the ratio of $A/(Q \times t)$ and unit conversion constant (10^9).

The BAM-1020 operates on a one-hour cycle starting at the top of each hour with filter movement between each of the following: (1) measurements starting with an 8-minute beta measurement on a clean filter spot at the beginning of each hour; (2) a sampling period of 42 minutes; (3) an 8-minute beta measurement of a particle loaded filter spot; and (4) an explicit cycle taking a total of 58 minutes with the other 2 minutes of the hour used for tape and nozzle movements during the cycle. Figure D.1 shows a detailed picture of the BAM's sample head, filter tape, and beta source and detector.

FIGURE D.1

Picture of the Met One beta attenuation monitor unit showing the beta source and the filter tape used to collect the particulate matter samples

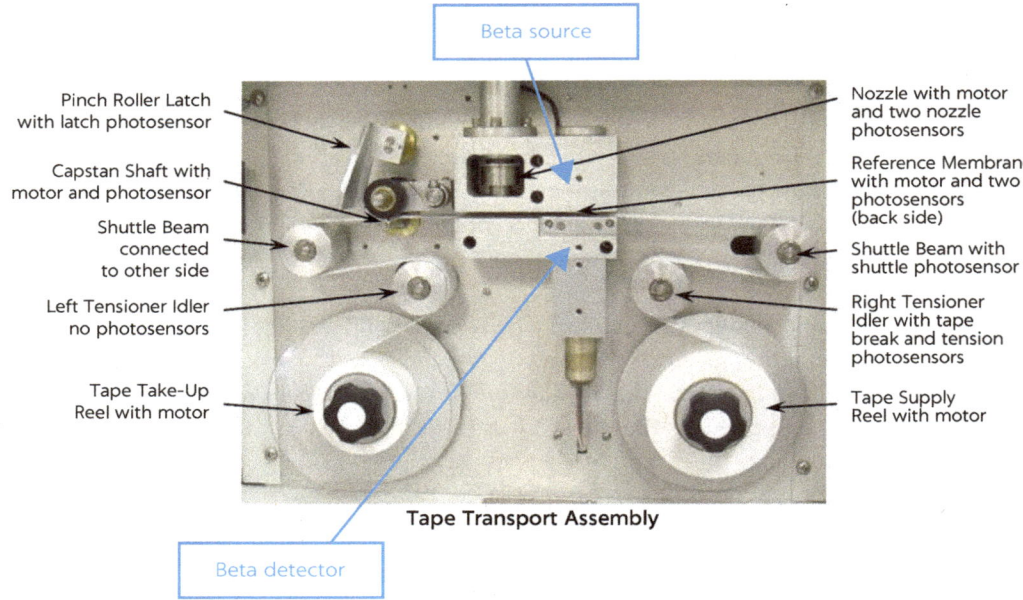

Measurement uncertainties in the BAM's $PM_{2.5}$ can be obtained from estimates of uncertainties, the equation for $PM_{2.5}$ mass concentrations given above, and standard error propagation. The random (precision) uncertainties in the BAM's $PM_{2.5}$ calculation include (1) flow rate, sample time, spot size (minor, about 2 percent); (2) calibration (μ) and span values (10 percent); (3) counting statistics (I and I_o) (about 2.5 micrograms per square meter); and (4) baseline and electronic drift (minor, about 2 percent). Other uncertainties and systematic bias are more difficult to assess because they are operationally dependent on relative humidity changes during sampling, site-location factors, operator error, and volatile PM components that may evaporate from the filter during sampling. The manufacturer states an uncertainty (σ) of 2.4 micrograms per cubic meter (variance of 72 one-hour averages of zero particle measurements). Thus, any single hourly measurement should fall within $3\sigma = 7.2$ micrograms per cubic meter of the expected value.

OPERATING METHODOLOGIES AND UNCERTAINTIES ASSOCIATED WITH THE TAPERED ELEMENT OSCILLATING MICROBALANCE TECHNIQUE

Tapered element oscillating microbalance (TEOM) measurement principle: Sample air is drawn through a filter cartridge (Teflon-coated glass fiber) mounted on the tip of a hollow glass tube (the tapered element) at 3 liters per minute (split from a total 16.67 liters per minute flow through a size-cut sampling inlet; figure D.2). The base of the tube cannot move, but the tip is free to vibrate at its natural frequency. Thus, the tube acts as a tuning fork or a hollow cantilever beam with an associated spring rate and mass. Any additional weight from particles that collect on the filter changes the frequency at which the tube oscillates. The electronic circuitry senses this change and calculates the particle mass rate from the magnitude of the frequency change. The instrument then returns the vibrating glass tube to its natural frequency ready for the next measurement. The TEOM maintains a constant temperature and flow rate and electronically smooths the readings to reduce noise. Dividing the mass rate by the flow rate provides a continuous output of the particle mass concentration.

Measurement of mass: The frequency of a spring-mass system follows the equation

$$f = \left(\frac{K}{M}\right)^{0.5}, \tag{D.3}$$

where f is the frequency of oscillation of the tapered element with the particle filter, K is the spring constant, and M is the mass of the tapered element.

To measure the particulate matter (PM) collected on the particle filter, differences in mass are obtained by

$$dm = K_0 \left(\frac{1}{f_1^2} - \frac{1}{f_0^2}\right), \tag{D.4}$$

where dm is the change in mass of the tapered element due to collected PM, K_0 is the spring constant and unit conversion constants (calibrated using known masses), and f_0 and f_1 are the initial and final frequencies (hertz) for a measurement.

FIGURE D.2
Schematic diagram of a tapered element oscillating microbalance

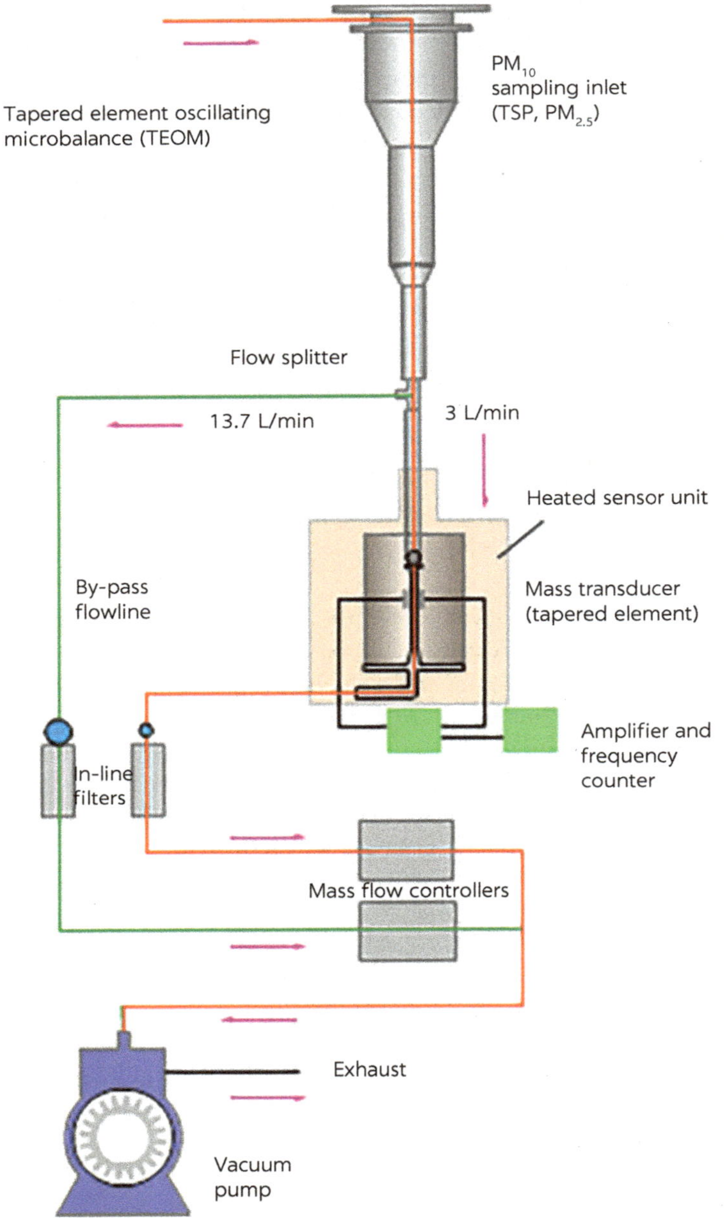

Source: © The State of Queensland 2021 (CC BY 4.0). https://www.qld.gov.au/environment/pollution/monitoring/air-pollution/oscillating-microbalance.
Note: L/min = liters per minute; PM = particulate matter; TSP = total suspended particulate.

The ambient concentration of particulate matter (micrograms per cubic meter) is then

$$C\left(\mu g/m^3\right) = \frac{dm\left(\mu g\right)}{Q\left(m^3/s\right) \times t\left(s\right)},$$

(D.5)

where C is the ambient PM concentration (micrograms per cubic meter), dm is the change in mass of the tapered element due to collected PM (micrograms),

Q is the rate at which ambient air is being sampled (cubic meters per second), and t is the sampling time (seconds).

TEOM measurement uncertainties:

$$C\left(\mu g / m^{3}\right)=\frac{dm}{(Q \times t)} \qquad (D.6)$$

Random (precision):

- Flow rate, (minor, about 2 percent)
- Calibration (K_0) and span values (10 percent)
- Baseline and electronic drift (minor, about 2 percent)
- Noise due to difference measurements (significant at short time frames)

Systematic (bias):

- RH dependence (addressed via sample heating)
- Environmental vibrations (affect stability of oscillations)
- Volatile PM (specifically nitrates and semivolatile organics)
- Operator error

Uncertainties typically are combined via "root-sum-of-squares":

$$\sigma^{2}(C)=\sum \frac{\partial C}{\partial x_{i}} \sigma^{2}\left(x_{i}\right)+2 \sum \sum \frac{\partial C}{\partial x_{i}} \frac{\partial C}{\partial x_{j}} \sigma\left(x_{i}, x_{j}\right). \qquad (D.7)$$

The manufacturer states precision uncertainty of ±2.0 micrograms per cubic meter for a one-hour average:

- If this is a single σ precision, any single measurement should fall within 3σ = 6.0 micrograms per cubic meter of the expected value
- The TEOM stated accuracy uncertainty of ±0.75 percent for a one-hour average on the mass measurement (dm)
- If this is a single σ accuracy, any single one-hour measurement should fall within 3σ = 2.25 percent of the expected value (this needs to be combined with flow rate–measurement uncertainties)
- Thus, TEOM measurement uncertainty (for a well-operated system under the Environmental Protection Agency protocols and requirements) would be the greater of these uncertainties
- The unknown factors include site location, volatile PM losses, and operator errors.

There are several TEOM base models (1400a, 1400b, 1405) and methods (standard and FDMS). The base TEOM instrument is the same instrument detection scheme using the same tapered element oscillating microbalance. The various TEOM model numbers refer to the following:

- Model 1400a: Sampling with $PM_{2.5}$ cyclone
- Model 1400b: Sampling with PM_{10} cyclone
- Model 1405: Sampling with a virtual impactor that separates particles into fine ($PM_{2.5}$) and coarse particle modes (PM_{10}–$PM_{2.5}$)
- Models with –F or –DF: include a Filter Dynamics Measurement System (FDMS) upstream of the TEOM (see schematic in figure D.3).

The TEOM 1405-DF Dichotomous Ambient Particulate Monitor with FDMS is designated as federal equivalent method EQPM-0609-182 for $PM_{2.5}$. In this

FIGURE D.3

Tapered element oscillating microbalance–filter dynamics measurement system schematic

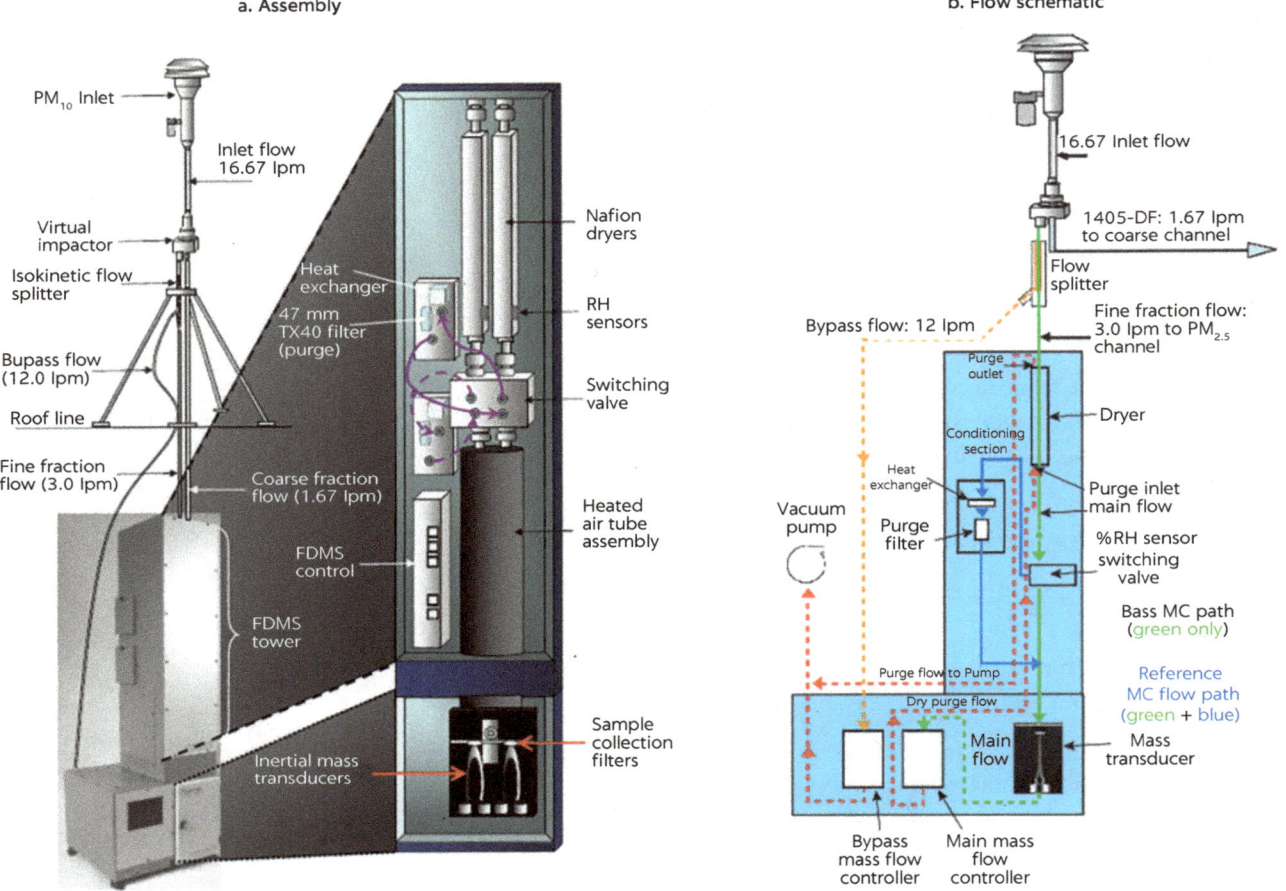

Source: Standard Operating Procedure for the Continuous Measurement of Particulate Matter Thermo Scientific TEOM® 1405-DF Dichotomous Ambient Particulate Monitor with FDMS® Federal Equivalent Method EQPM-0609-182 for PM$_{2.5}$ STI-905505.03-3657-SOP, available at https://www3.epa.gov /ttnamti1/files/ambient/pm25/sop_project/905505_TEOM_SOP_Draft_Final_Sept09.pdf (pages 3-3 and 3-4).
Note: FDMS = Filter Dynamics Measurement System; lmp = liters per minute; MC = mass concentration; mm = millimeters; PM = particulate matter.

version of the instrument, all sample flows are passed through a diffusion dryer to remove water vapor. Every six minutes, a switching valve alternates the sample flows between the base- and reference-sample periods. During the base period, the PM sample is collected normally, and the differential mass is measured at 30° Celsius (C). At this temperature, volatile PM will possibly evaporate and not be measured. During the reference period, the sample flow is diverted through a chilled filter (4°C to 10°C) to remove and retain both the nonvolatile and volatile PM prior to passing through the particle filter of the TEOM. Thus, the reference period is when dry, particle-free air is passed through the particle filter of the TEOM. The "PM" mass measurement during this period is a differential mass measurement that is either near zero or negative, providing an approximate measure of the amount of volatile PM that evaporates from the TEOM particle filter. Ambient PM concentrations are calculated from base-reference mass measurements. Logging parameters, recommended maintenance, and calibration details are shown below:[1]

- **1405-F/DF** TEOM recommended logging parameters:[2]
  - FEM MC
  - $PM_{2.5}$ Base MC
  - $PM_{2.5}$ Ref MC
  - **PMc MC**
  - **PMc Base MC**
  - **PMc Ref MC**
  - Ambient temperature
  - Ambient humidity
  - Ambient pressure
  - Vacuum pump pressure
  - Status
  - TEOM A filter loading
  - TEOM A dryer temperature
  - TEOM A dryer dew point
  - **TEOM B filter loading**
  - **TEOM B dryer temperature**
  - **TEOM B dryer dew point**
  - Three additional available for user selection (20 total)

- Recommended maintenance intervals:
  - Replace TEOM filter: Filter loading nears 75 percent or every 30 days
  - Replace chilled filter: When you exchange the TEOM filter
  - Clean PM_{10} inlet: With every TEOM filter exchange
  - Clean virtual impactor (1405-DF): With every TEOM filter exchange
  - Replace in-line filter: Every 6 months
  - Clean coolers: Once a year
  - Clean switching valve: Once a year
  - Clean air inlet system: Once a year
  - Rebuild vacuum pump: 12 to 18 months
  - Dryer refurbishment: Once a year

- Recommended interval of frequency:
  - Ambient temperature: Audit monthly/calibration yearly
  - Ambient pressure: Audit monthly/calibration yearly
  - Flow ($PM_{2.5}$, coarse, bypass): Audit monthly/calibration yearly
  - Leak check: Monthly
  - Analog outputs: Once a year
  - Mass transducer: Audit once a Year

INTERPRETING GROUND-LEVEL $PM_{2.5}$ DATA FROM LOW- AND MIDDLE-INCOME COUNTRIES WITH LIMITED METADATA

In many low- and middle-income countries (LMICs), there are organizations, researchers, scientists, and members of the general public who are keenly interested in understanding *their* local air quality, especially if air pollution episodes occur with any regularity and in turn directly affect the way people feel. Nearly all publicly accessible information on real-time air pollution concentrations is just that—a single data point, often a $PM_{2.5}$ mass concentration. Little to no metadata are available with which to draw additional concrete expectations for instrument performance, accuracy, precision, or reliability. In such situations, one way to evaluate the legitimacy of the reported $PM_{2.5}$ levels is to compare measurements from different monitoring stations across a given area of interest. More populous LMICs often have multiple government-run air-quality monitoring stations spread across each city. In recent years, US Diplomatic Posts throughout these same LMICs have established a $PM_{2.5}$ monitoring capacity (equipped with BAM instruments) on embassy roofs. Accessible data from local and US government sites within the same city present the opportunity to evaluate the consistency of reported $PM_{2.5}$ concentrations in the area. This concept is demonstrated below by examining seasonal, monthly, weekly, and diurnal trends in $PM_{2.5}$ data across Delhi, India, and Lima, Peru, throughout the years 2016–17.

Delhi, India

Map D.1 reveals a shaded area of about 50 square miles, the perimeter of which is defined with air-quality monitoring sites in Delhi. Site 1 corresponds to the location of the US Diplomatic Post, and sites 2–7 are government-run monitoring stations equipped to measure $PM_{2.5}$ mass concentrations. Working with the OpenAQ online database, $PM_{2.5}$ data measured in 2016 and 2017 were retrieved for each of the stations shown on the map. The series of plots displayed below show the time-series, seasonal, monthly, weekly, and diurnal profiles in $PM_{2.5}$ for each monitoring location. In this metadata "blind" scenario, the US Diplomatic Post represents slightly higher reliability than the local government stations based on the fact that embassy staff employ the same standard operating procedure as recommended by the EPA to keep the BAM system maintained and in

MAP D.1

Proximity and distribution of government-run air-quality monitoring stations around the US Diplomatic Post in Delhi, India

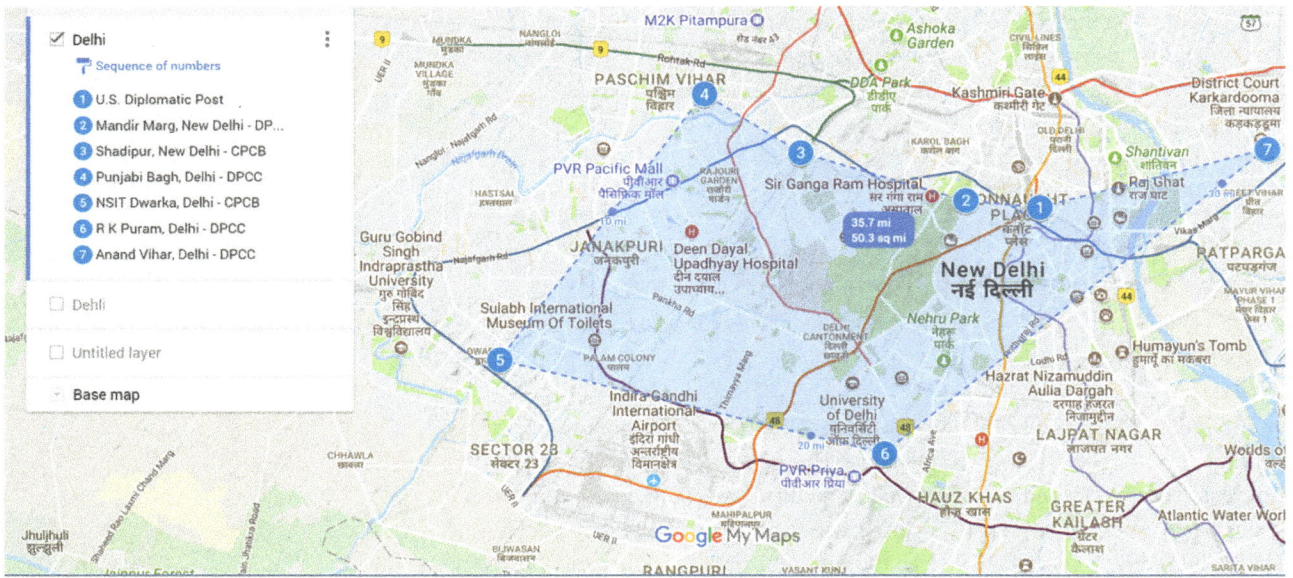

Source: World Bank, produced using Google Maps.
Note: CPCB = Central Pollution Control Board; DPCC = Delhi Pollution Control Committee.

good working order. These same assurances are not necessarily present in the local government data feeds, and thus the direct head-to-head comparisons with embassy data can be a useful tool for validating the government $PM_{2.5}$ trends. The top panel of the analysis frame (figure D.4) shows the $PM_{2.5}$ time series, season-average $PM_{2.5}$, monthly average $PM_{2.5}$, and weekly average $PM_{2.5}$. The lower panel in each analysis frame breaks down the $PM_{2.5}$ trends further, revealing diurnal (hourly) profiles for the overall data set and then each of the four seasons. One should take careful notice of not only the relative trends observed at each location but also the magnitudes of the peaks and valleys in $PM_{2.5}$. In each series of graphs, the $PM_{2.5}$ concentration axes (all y-axes in the graph series) are normalized to the maximum observed concentration (or error bar) in that graph.

The embassy BAM data reveal the strong seasonal dependence of Delhi's $PM_{2.5}$ concentrations, with extremely high (greater than 200 micrograms per cubic meter) levels observed between November and January. There is virtually zero weekday versus weekend difference in $PM_{2.5}$ loadings, and hourly diurnal profiles demonstrate the pronounced influence of boundary layer dynamics comingled with pollution source intensity and activity profiles: higher $PM_{2.5}$ concentrations overnight and into the morning hours when the boundary layer is lowest with a significant decrease in pollutant concentration in the mid-to-late afternoon as the boundary layer rises, diluting the $PM_{2.5}$ levels. To provide an empirical basis to assess the consistency of ground-level $PM_{2.5}$ measurements throughout Delhi, the same series of graphs are plotted in figures D.5 through D.10. Each figure corresponds to a different government-run air-quality monitoring location (marked as 2–7 on map D.1). Generally, the observed seasonal, monthly, and weekly trends in $PM_{2.5}$ levels from the government stations are consistent with the $PM_{2.5}$ data acquired at the embassy. There are some

FIGURE D.4

Time-series, seasonal, monthly, weekly, and diurnal $PM_{2.5}$ mass concentration data measured from the US Diplomatic Post in Delhi, India

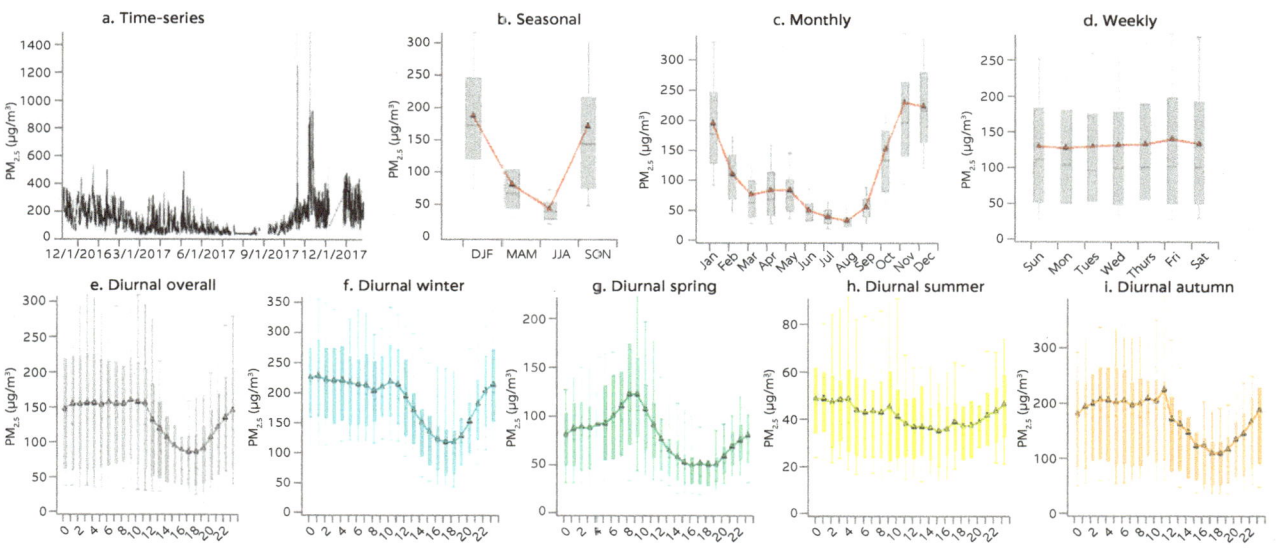

Source: World Bank.
Note: DJF = December, January, February; MAM = March, April, May; JJA = June, July, August; SON = September, October, November. $PM_{2.5}$ = particulate matter with an aerodynamic diameter less than or equal to 2.5 microns; μg/m³ = micrograms per cubic meter.

FIGURE D.5

PM₂.₅ trends from data acquired at the Mandir Marg, New Delhi, India–DPCC monitoring site, 1.45 miles west of US Diplomatic Post

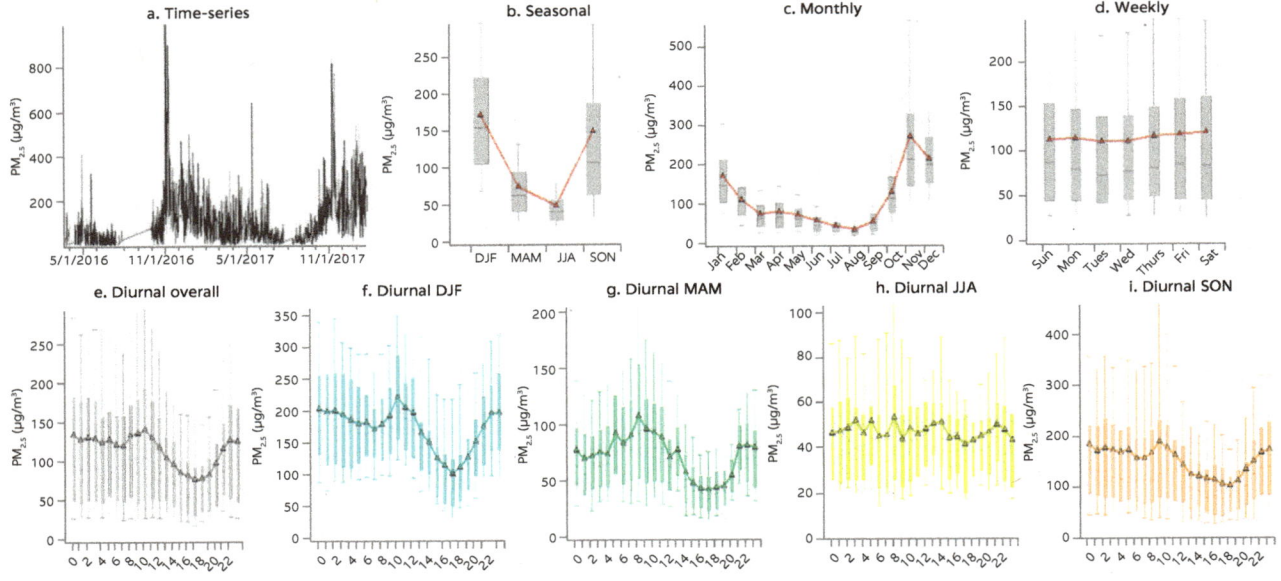

Source: World Bank.

Note: DJF = December, January, February; MAM = March, April, May; JJA = June, July, August; SON = September, October, November. DPCC = Delhi Pollution Control Committee; PM₂.₅ = particulate matter with an aerodynamic diameter less than or equal to 2.5 microns; µg/m³ = micrograms per cubic meter.

FIGURE D.6

PM₂.₅ trends from data acquired at the Shadipur, New Delhi, India–CPCB monitoring site, 4.75 miles northwest of US Diplomatic Post

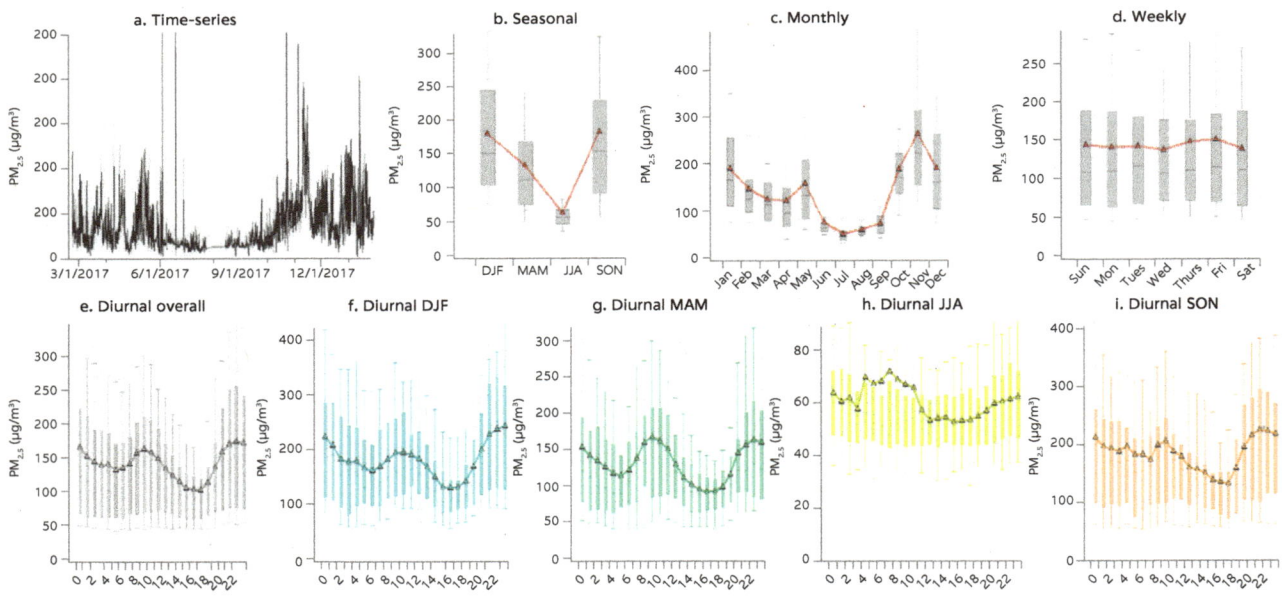

Source: World Bank.

Note: DJF = December, January, February; MAM = March, April, May; JJA = June, July, August; SON = September, October, November. CPCB = Central Pollution Control Board; PM₂.₅ = particulate matter with an aerodynamic diameter less than or equal to 2.5 microns; µg/m³ = micrograms per cubic meter.

FIGURE D.7

PM~2.5~ trends from data acquired at the Punjabi Bagh, Delhi, India–DPCC monitoring site, 6.91 miles northwest of US Diplomatic Post

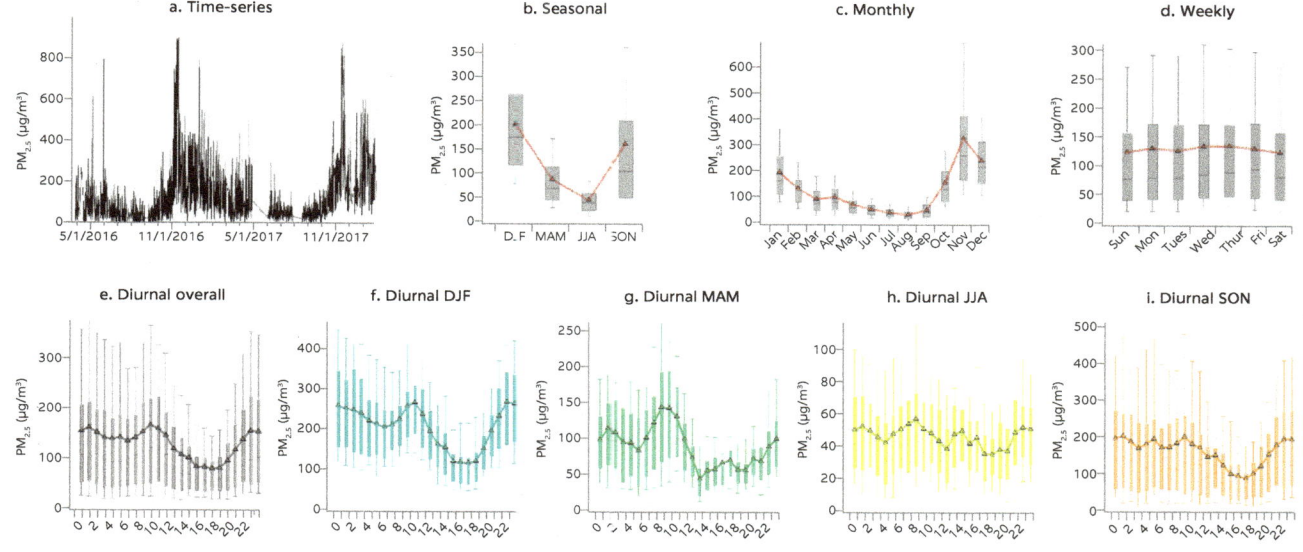

Source: World Bank.

Note: DJF = December, January, February; MAM = March, April, May; JJA = June, July, August; SON = September, October, November. DPCC = Delhi Pollution Control Committee; PM~2.5~ = particulate matter with an aerodynamic diameter less than or equal to 2.5 microns; μg/m³ = micrograms per cubic meter.

FIGURE D.8

PM~2.5~ trends from data acquired at the NSIT Dwarka, Delhi, India–CPCB monitoring site, 11 miles southwest of US Diplomatic Post

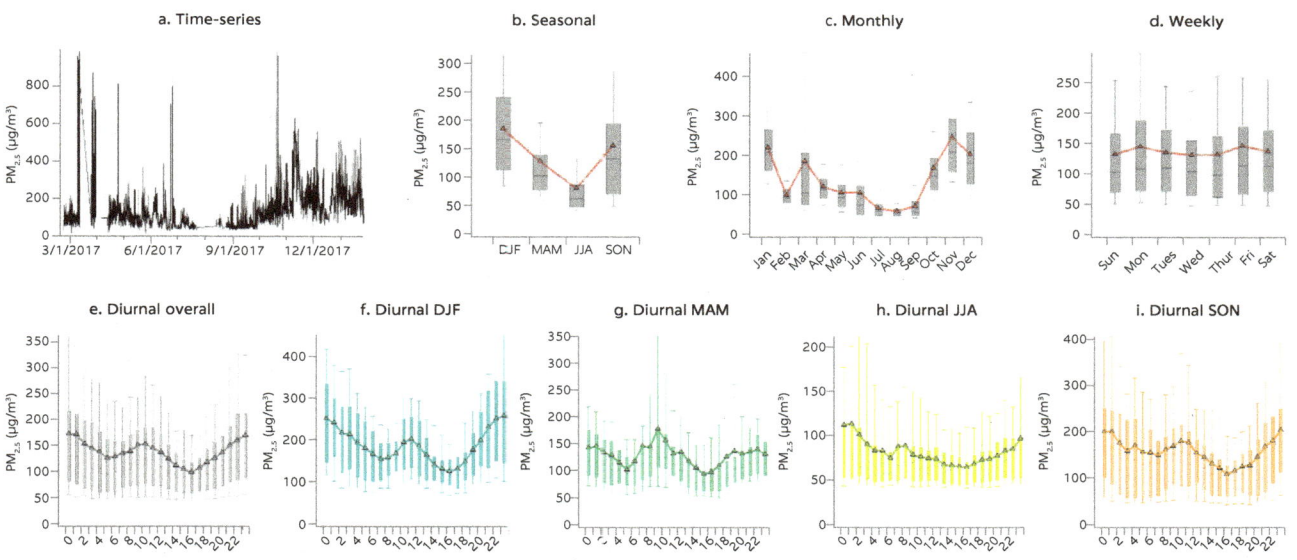

Source: World Bank.

Note: DJF = December, January, February; MAM = March, April, May; JJA = June, July, August; SON = September, October, November. CPCB = Central Pollution Control Board; NSIT = Netaji Subhas Institute of Technology; PM~2.5~ = particulate matter with an aerodynamic diameter less than or equal to 2.5 microns; μg/m³ = micrograms per cubic meter.

FIGURE D.9

PM$_{2.5}$ trends from data acquired at the Ramakrishna Puram, Delhi, India–DPCC monitoring site, 5.69 miles south-southwest of US Diplomatic Post

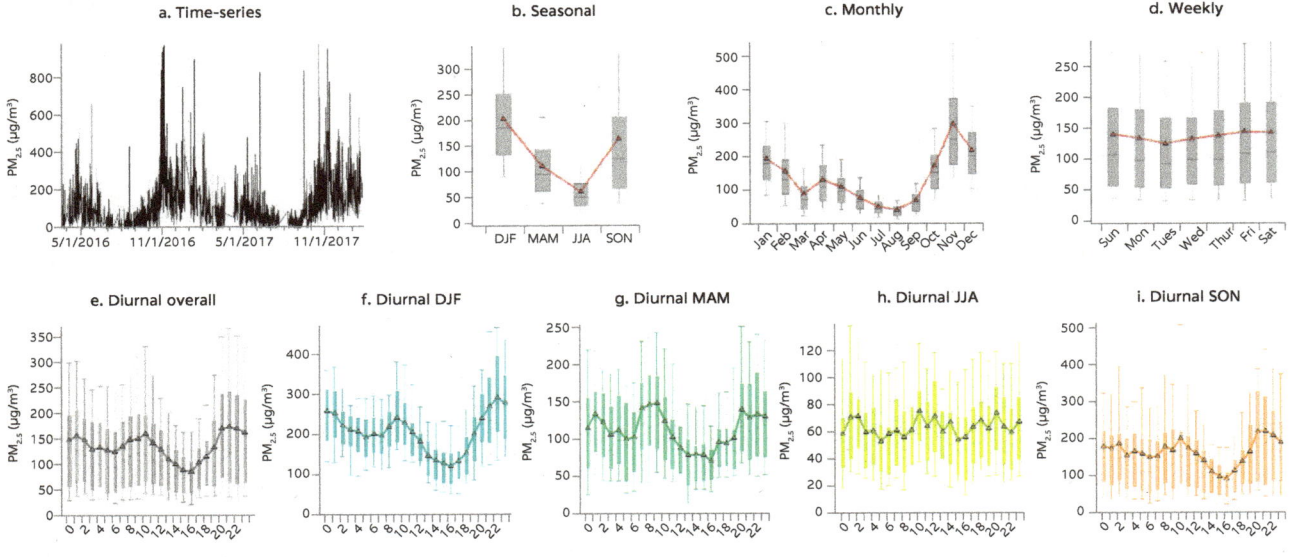

Source: World Bank.
Note: DJF = December, January, February; MAM = March, April, May; JJA = June, July, August; SON = September, October, November. DPCC = Delhi Pollution Control Committee; PM$_{2.5}$ = particulate matter with an aerodynamic diameter less than or equal to 2.5 microns; μg/m³ = micrograms per cubic meter.

FIGURE D.10

PM$_{2.5}$ trends from data acquired at the Anand Vihar, Delhi, India–DPCC monitoring site, 4.6 miles east-northeast of US Diplomatic Post

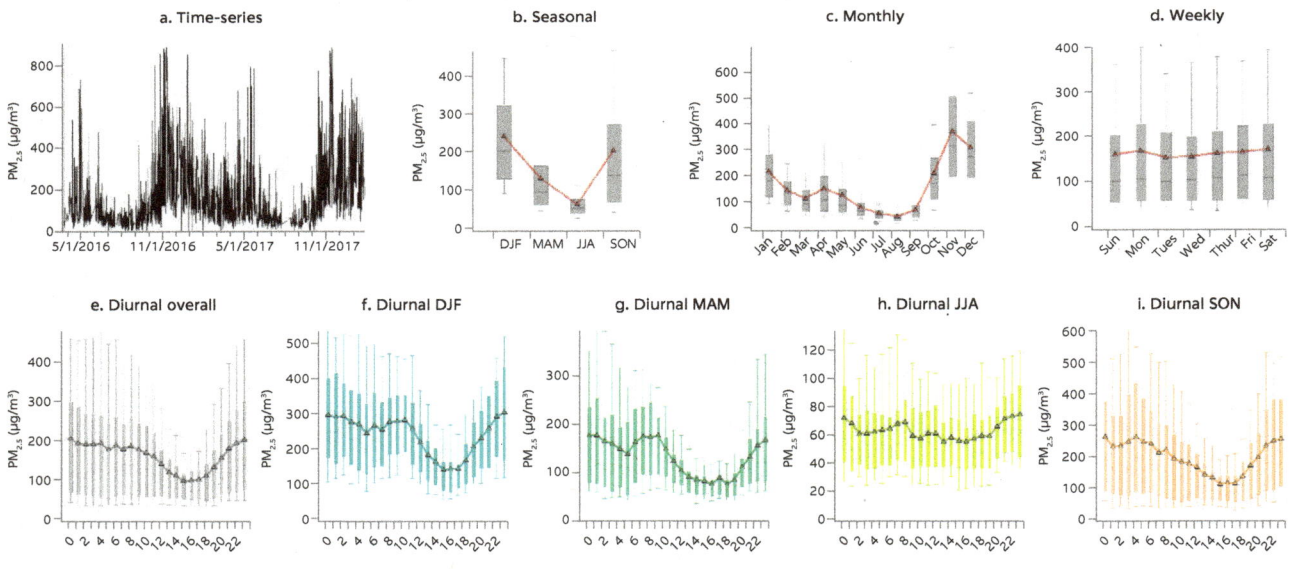

Source: World Bank.
Note: DJF = December, January, February; MAM = March, April, May; JJA = June, July, August; SON = September, October, November. DPCC = Delhi Pollution Control Committee; PM$_{2.5}$ = particulate matter with an aerodynamic diameter less than or equal to 2.5 microns; μg/m³ = micrograms per cubic meter.

differences in the reported data between stations, but no egregiously disparate values are found that would be indicative of instrument malfunction or data tampering across this subset of monitoring locations.

Lima, Peru

The embassy BAM data from Lima, Peru, reveal nearly an order of magnitude lower $PM_{2.5}$ concentrations than observed in Delhi, with a far less pronounced seasonal dependence. There is little to no weekday-versus-weekend difference in $PM_{2.5}$ loadings, and hourly diurnal profiles indicate that $PM_{2.5}$ pollution source intensity and activity profiles result in highest $PM_{2.5}$ concentrations in the midmorning hours (except for June, July, and August in the embassy data, when peak $PM_{2.5}$ concentrations shift to early afternoon). To provide an empirical basis to assess the consistency of ground-level $PM_{2.5}$ measurements throughout Lima, the same series of graphs are plotted in figures D.11 through D.15. Each figure corresponds to a different government-run air-quality monitoring location (marked as 2–5 on map D.2). Generally, the observed seasonal, monthly, and weekly trends in $PM_{2.5}$ levels from the government stations are reasonably consistent with the $PM_{2.5}$ data acquired at the embassy, but important differences are seen in the magnitude of the $PM_{2.5}$ concentrations reported at different locations. A moderately polluted area such as Lima highlights the need for accurate, reproducible measurements in a manner that is often lost in the much more heavily polluted LMICs (such as Delhi). The embassy data shown in figure D.12 also highlight the importance of having complete data sets over which the $PM_{2.5}$ trends can be analyzed. Missing data from the embassy site complicates the head-to-head intercomparisons somewhat, especially if the surrounding stations had more complete data sets (which is the case for most of the government locations examined here). Nevertheless, inferences can be drawn by completing head-to-head analyses focused on the times of year where data are available.

Year-round, the embassy $PM_{2.5}$ concentrations are about 30 ± 10 micrograms per cubic meter. In comparison, San Borja, Peru (2.7 miles southwest of the embassy) reports concentrations on average about 30 percent lower throughout the year, and Campo de Marte, Peru, reports concentrations about 50 percent lower than the embassy. In contrast, the other two stations, Santa Anita, Peru, and Ate, Peru, report concentrations about 10–30 percent higher than the embassy. In this way, the Lima data sets begin to reveal the complexity of combining $PM_{2.5}$ measurements from different agencies at different locations with minimal metadata. Closer inspection of the prevailing meteorology alongside the pollution source activity profiles and density within the city would further elucidate the extent to which the differences observed across the monitoring stations are derived from atmospheric conditions or instrument performance. The diurnal analyses presented here represent a first step toward building a more comprehensive approach to validating ground-level $PM_{2.5}$ concentrations in LMICs.

FIGURE D.11

Time-series, seasonal, monthly, weekly, and diurnal PM$_{2.5}$ mass concentration data measured from the US Diplomatic Post in Lima, Peru

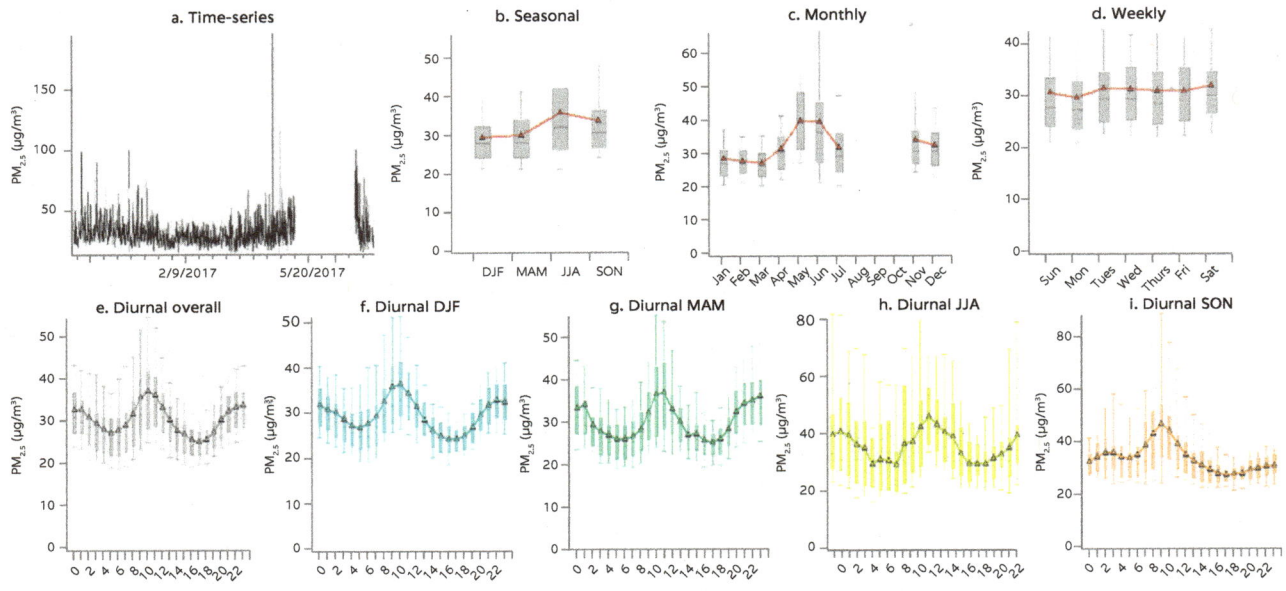

Source: World Bank.
Note: DJF = December, January, February; MAM = March, April, May; JJA = June, July, August; SON = September, October, November.
PM$_{2.5}$ = particulate matter with an aerodynamic diameter less than or equal to 2.5 microns; µg/m³ = micrograms per cubic meter.

FIGURE D.12

PM$_{2.5}$ trends from data acquired at the San Borja, Peru, monitoring site, 2.7 miles southwest of US Diplomatic Post

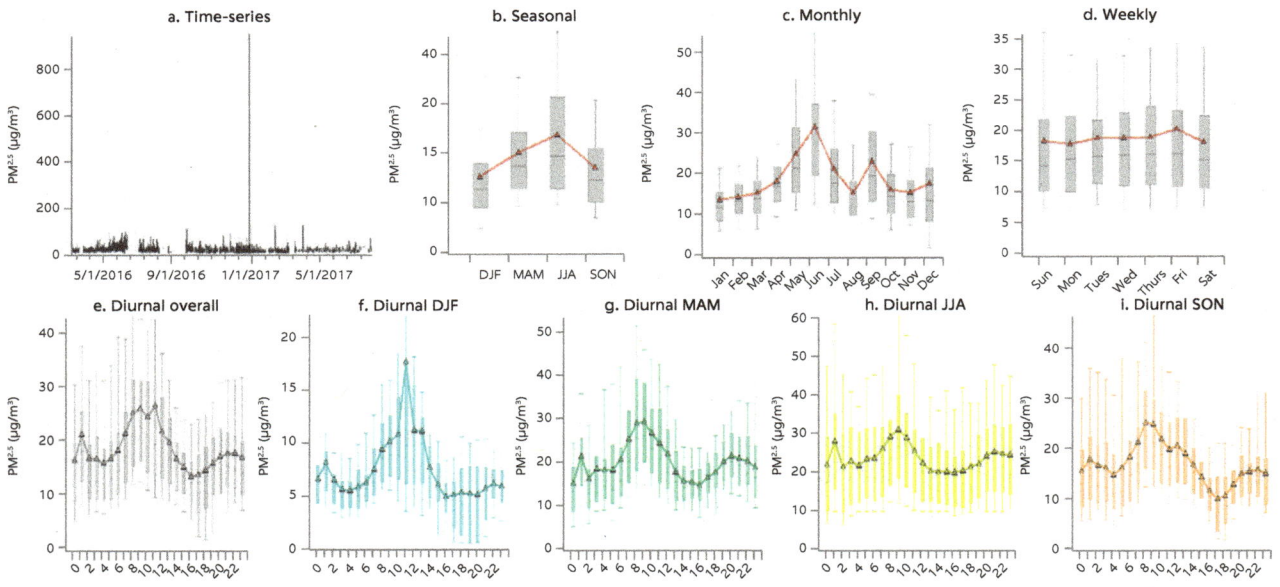

Source: World Bank.
Note: DJF = December, January, February; MAM = March, April, May; JJA = June, July, August; SON = September, October, November.
PM$_{2.5}$ = particulate matter with an aerodynamic diameter less than or equal to 2.5 microns; µg/m³ = micrograms per cubic meter.

FIGURE D.13

PM$_{2.5}$ trends from data acquired at the Campo de Marte, Peru, monitoring site, 5.4 miles northwest of US Diplomatic Post

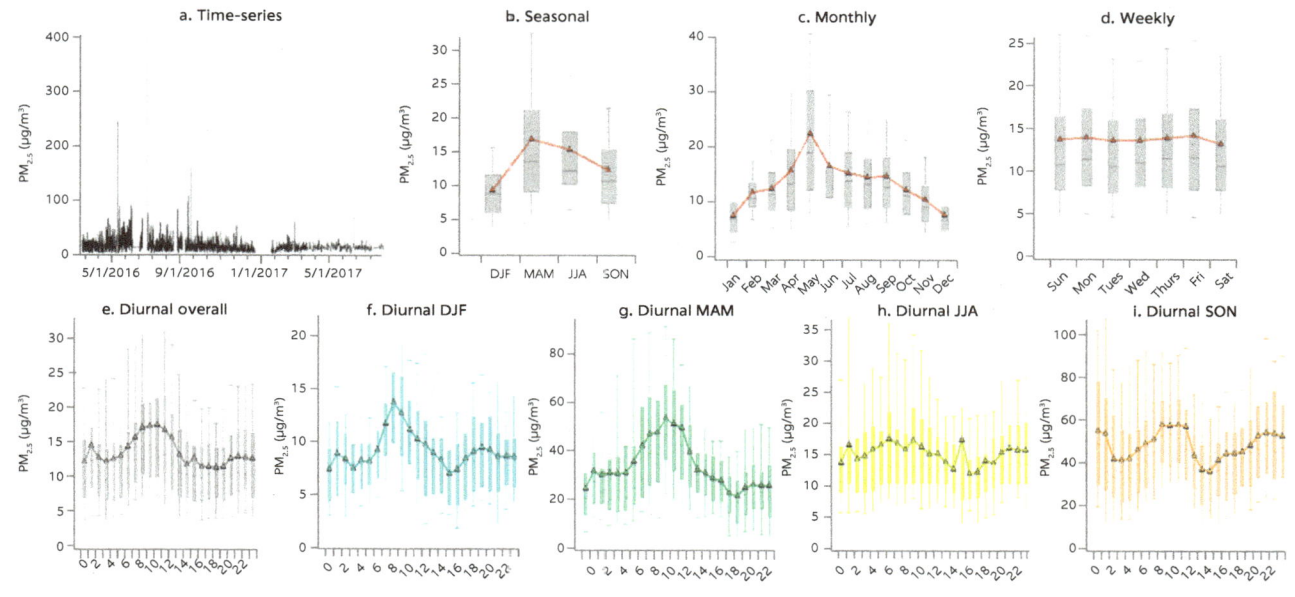

Source: World Bank.

Note: DJF = December, January, February; MAM = March, April, May; JJA = June, July, August; SON = September, October, November. PM$_{2.5}$ = particulate matter with an aerodynamic diameter less than or equal to 2.5 microns; μg/m³ = micrograms per cubic meter.

FIGURE D.14

PM$_{2.5}$ trends from data acquired at the Santa Anita, Peru, monitoring site, 3.9 miles north of US Diplomatic Post

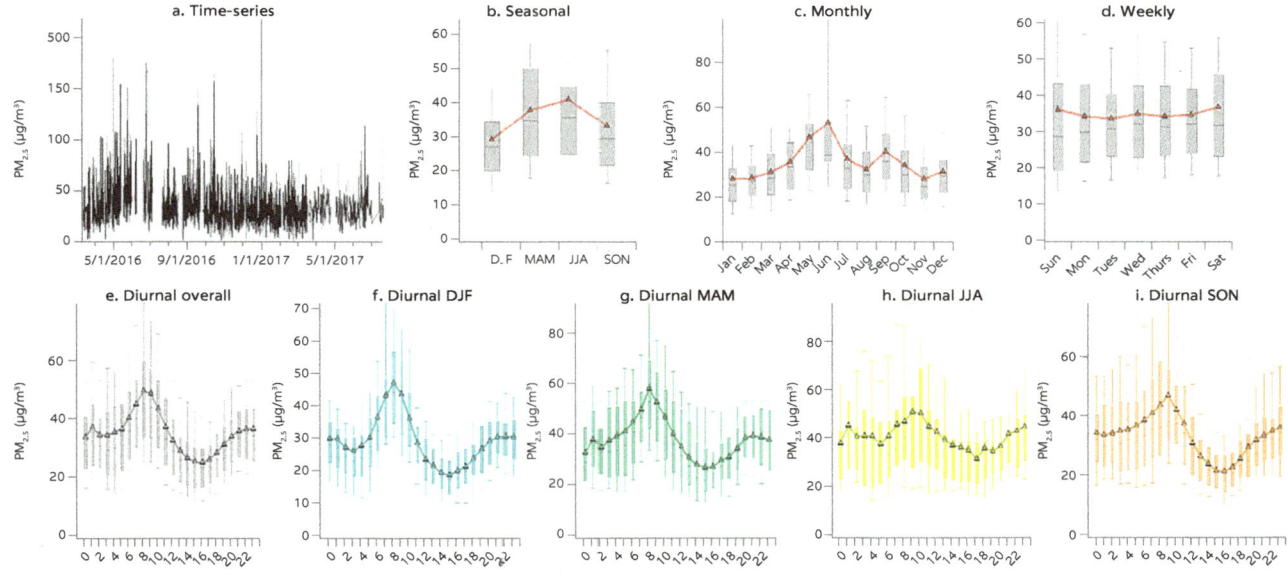

Source: World Bank.

Note: DJF = December, January, February; MAM = March, April, May; JJA = June, July, August; SON = September, October, November. PM$_{2.5}$ = particulate matter with an aerodynamic diameter less than or equal to 2.5 microns; μg/m³ = micrograms per cubic meter.

FIGURE D.15

PM$_{2.5}$ trends from data acquired at the Ate, Peru, monitoring site, 6.1 miles northeast of US Diplomatic Post

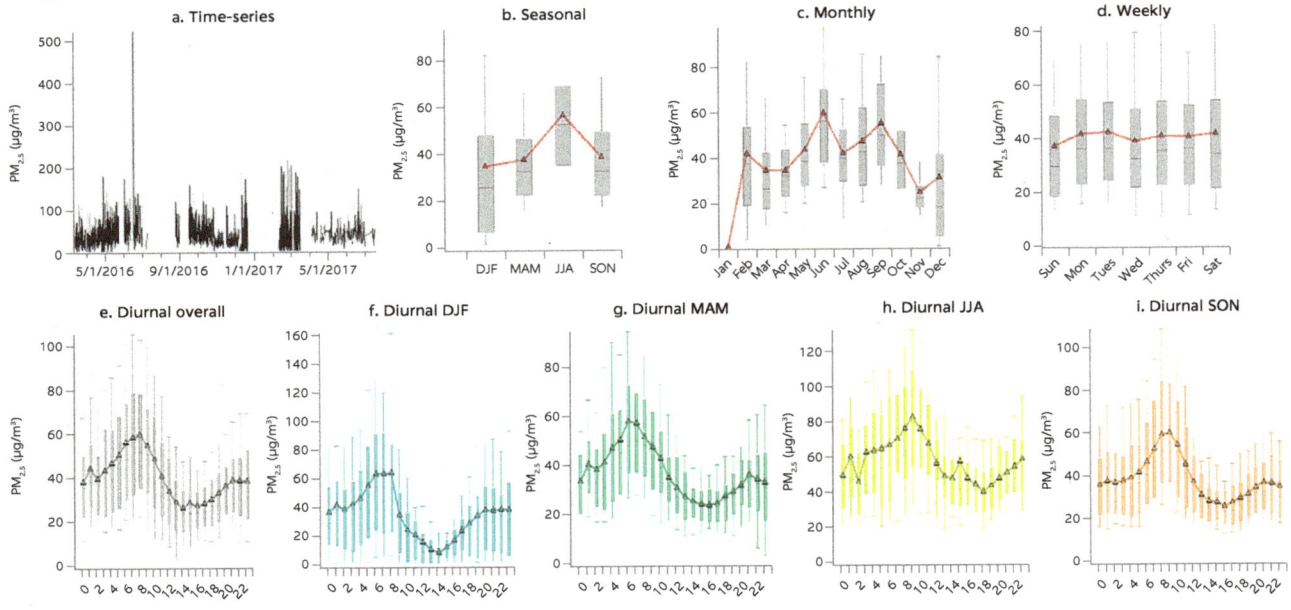

Source: World Bank.
Note: DJF = December, January, February; MAM = March, April, May; JJA = June, July, August; SON = September, October, November.
PM$_{2.5}$ = particulate matter with an aerodynamic diameter less than or equal to 2.5 microns; μg/m³ = micrograms per cubic meter.

MAP D.2

Illustration of the proximity of the US Diplomatic Post in Lima, Peru, relative to a subset of the nearest government-run assurance quality monitoring stations for which OpenAQ PM$_{2.5}$ data were available

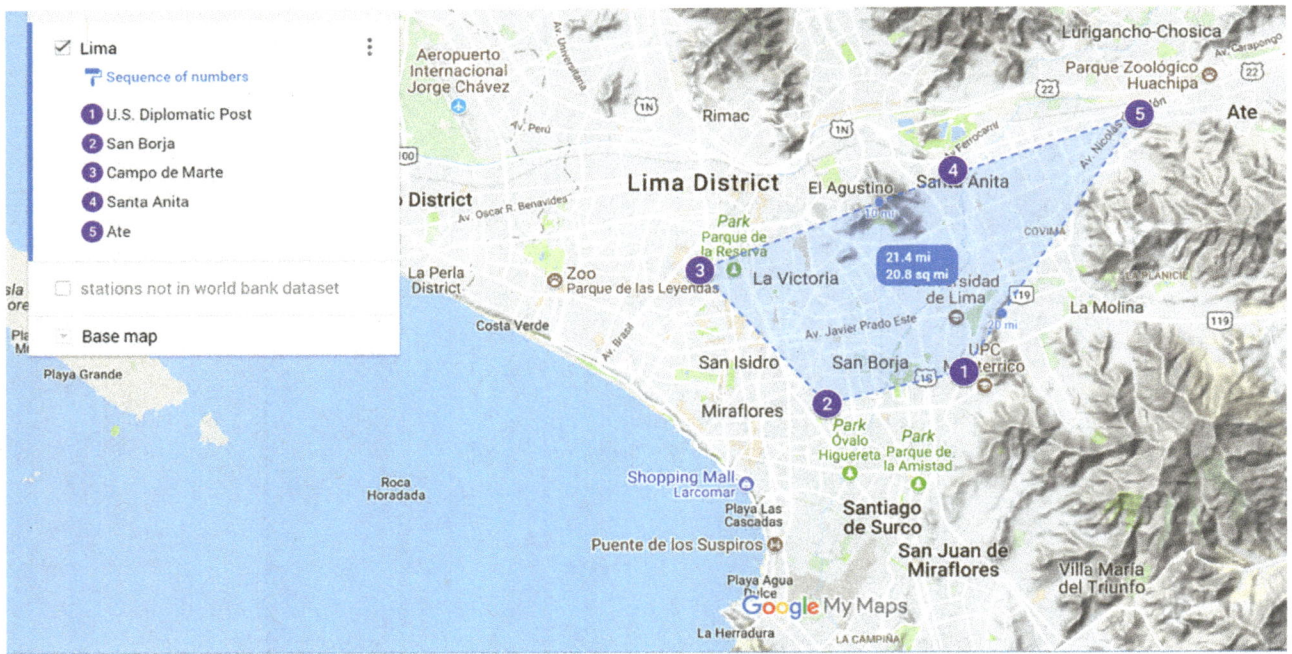

Source: World Bank, produced using Google Maps.
Note: Government assurance quality stations ranged between about 2.5 to 6 miles from the US Diplomatic Post, covering an area of about 20.8 square miles (blue shaded area). OpenAQ = openaq.org; PM$_{2.5}$ = particulate matter with an aerodynamic diameter less than or equal to 2.5 microns.

NOTES

1. https://www3.epa.gov/ttnamti1/files/ambient/pm25/sop_project/905505_TEOM_SOP _Draft_Final_Sept09.pdf.
2. Bold text represents additional parameters logged for the 1045-DF TEOM; other text represents the parameters logged for both the 1045-F and 1045-DF TEOM.